LASERS AND ELECTRO-OPTICS RESEARCH AND TECHNOLOGY

DIGITAL SIGNAL PROCESSING (DSP)

FUNDAMENTALS, TECHNIQUES AND APPLICATIONS

Lasers and Electro-Optics Research and Technology

Additional books in this series can be found on Nova's website under the Series tab.

Additional e-books in this series can be found on Nova's website under the e-book tab.

LASERS AND ELECTRO-OPTICS RESEARCH AND TECHNOLOGY

DIGITAL SIGNAL PROCESSING (DSP)

FUNDAMENTALS, TECHNIQUES AND APPLICATIONS

JUAN ZHANG
EDITOR

nova publishers
New York

Library of Congress Cataloging-in-Publication Data

Names: Zhang, Juan (Engineer), editor.
Title: Digital signal processing (DSP) : fundamentals, techniques and applications / [edited by] Juan Zhang (Key Laboratory of Specialty Fiber Optics and Optical Access Networks, School of Communication and Information Engineering, Shanghai University, China).
Description: Hauppauge, New York : Nova Science Publishers, Inc., [2016] | Series: Lasers and electro-optics research and technology | Includes bibliographical references and index.
Identifiers: LCCN 2016017058 (print) | LCCN 2016022541 (ebook) | ISBN 9781634851688 (hardcover) | ISBN 9781634852104 ()
Subjects: LCSH: Signal processing--Digital techniques.
Classification: LCC TK5102.9 .D532 2016 (print) | LCC TK5102.9 (ebook) | DDC 621.382/2--dc23
LC record available at https://lccn.loc.gov/2016017058

Published by Nova Science Publishers, Inc. † New York

CONTENTS

PREFACE

Digital signal processing (DSP) is the numerical manipulation (including measuring, filtering, producing or compressing) of signals.

The theoretical basis of DSP is the use of discrete domain signals (such as time, frequency and so on), which can be easily modeled, analyzed and processed digitally. So DSP algorithms have long been run on standard computers (such as with MATLAB), as well as on specialized processors such as field-programmable gate arrays (FPGAs).

Rapid and massive advances in DSP technology have been achieved over the past several decades. DSP technology revolutionized the electronics and even opto-electronic industries. The main applications of it include audio and speech signal processing, image processing, spectral signal processing, biomedical signal processing, seismic data processing, and so on. Optical communication plays a significant and increasing role in our society. Use of well-developed DSP techniques and algorithms to design the wavelength division multiplexing (WDM) devices is a wise use of existing technology. The interdisciplinary cooperation between DSP and optics will be crucial for the next-generation optical communications and integrated photonics.

This book covers a wide range of elementary and advanced topics in DSP, focusing on such areas as filter design algorithms, hardware/software techniques, and their applications. Subjects dealt with include the generalized distribution function, Mittag-Leffler expansion, active contour models, theory and design of the average filter, regularized bandpass filter and digital tracking filter, description and design of photonic bandgap structure and optical filters, implementation of low-pass, high-pass and band-pass filters using FPGAs. The last chapter introduces the DSP theory and hardware for obscured object identification, and its applications in the intelligent baggage scanners.

I hope this up-to-date book on DSP fundamentals, techniques and applications can fill some needs of researchers, designers and practicing engineers of software, hardware, devices, and systems.

Juan Zhang
Dec. 28, 2015

In: Digital Signal Processing (DSP)
Editor: Juan Zhang

ISBN: 978-1-63485-168-8

Chapter 1

GENERALIZED DISTRIBUTIONS, MITTAG-LEFFLER EXPANSION AND SAMPLING DISCONTINUOUS SIGNALS

Michael J. Corinthios*
École Polytechnique de Montréal, Université de Montréal,
Montréal, QC, Canada

ABSTRACT

Recently proposed generalized functions of a complex variable extend the domains of existence of Laplace and z-transform. In this chapter, basic properties of generalized distributions are extended. In particular, properties of generalized distributions in the context of sampling functions containing discontinuities are explored. Laplace and z-transforms of one-sided, two-sided periodic and exponentially modulated periodic impulses, hitherto nonexistent, and of which the Fourier transform does not exist, are evaluated using generalized distributions. Applications to the formulation of the sampling theorem are explored. The generalized distributions are shown to reveal an unusual decades-old anomaly in the well-known digital filtering approach of impulse invariance. The anomaly arises when Laplace and z-transform spectra are compared with the Mittag-Leffler Expansion. It is shown that the transformation, as is presently applied, does not produce the stated Fourier spectrum of the sampled signal. In fact it produces far more spectral aliasing than claimed instead of the minimum desired. In converting an analog filter of even a small order to a digital filter, an appreciable spectral deviation error is produced. Matlab still uses the erroneous approach. A new approach to impulse invariance, eliminating the error, is proposed. The general results have been confirmed in a recent paper by the author using regular distributions.

Keywords: digital signal processing, distributions theory, generalized distributions, Shannon's sampling theorem, Mittag-Leffler expansion, impulse invariance

* Email: michael.corinthios@polymtl.ca.

INTRODUCTION

Generalized functions have expanded considerably the domain of existence of the Fourier transform [1-10]. Weighted spectra leading to impulses on the complex Laplace and z-Transform planes have been proposed for the exponential decomposition of finite duration signals [11, 12]. The decomposition of infinite duration generally complex exponential continuous-time and discrete-time signals leads in general to diverging integrals and summations. Generalizing the Dirac-delta impulse has for objective to define transforms for a class of functions which leads to integrals that are not absolutely convergent. In this chapter, the distribution theoretic basis of the generalization is presented, followed by properties of the new distributions and the resulting bilateral transforms.

New relations governing the complex convolution theorem in the presence of generalized impulses are developed. Applications of the generalized distributions to the formulation of the sampling theorem and to the sampling of one-sided signals are explored. As an application, the digital signal processing approach for transforming continuous-time to discrete-time systems, and filters in particular, well covered in the literature for several past decades, is investigated in light of the obtained results. The generalized distributions are shown to reveal an unusual mathematical anomaly in the well-known digital filtering approach of impulse invariance. The anomaly arises when Laplace, *z*- and Fourier transform spectra are compared with results obtained by applying the Mittag-Leffler Expansion. It is shown that the transformation, as is presently applied, does not produce the stated Fourier spectrum of the sampled signal. In fact it produces more spectral aliasing than claimed, and not the minimum desired. In converting an analog filter of even a small order to a digital filter, an appreciable spectral deviation error is produced. A new approach to impulse invariance, eliminating the error, is proposed. The general results have been confirmed in a recent paper by the author using regular distributions.

GENERALIZED DISTRIBUTIONS IN LAPLACE DOMAIN

In the Laplace transform domain, a generalized distribution $G(s)$, function of the complex variable $s=\sigma+j\omega$, may be defined as an integral along a straight line contour in the s plane extending from a point $s=\sigma-j\infty$ to $s=\sigma+j\infty$ of the product of $G(s)$ with a test function $\Phi(s)$. For convenience we refer to this integral by the symbol $I_G[\Phi(s)]$, or simply $I_G[\Phi]$, and use the notation

$$I_G[\Phi(s)] = < G(s), \Phi(s) >_{\Re[s]=\sigma} = \int_{\sigma-j\infty}^{\sigma+j\infty} G(s)\Phi(s)\,ds \tag{1}$$

The test function $\Phi(s)$ has derivatives of any order along straight lines in the s plane going through the origin, and tends to zero more rapidly than any power of $|s|$. For example, if the generalized distribution is the generalized impulse $\xi(s)$ [13-16], we may write

$$I_G\left[\Phi(s)\right] = <\xi(s), \Phi(s)>_{\Re[s]=\sigma} = \int_{\sigma-j\infty}^{\sigma+j\infty} \xi(s)\Phi(s)\,ds = \begin{cases} j\Phi(0), & \sigma = 0 \\ 0, & ,\sigma \neq 0 \end{cases} \tag{2}$$

Basic Properties

In the following a selection of basic properties of generalized distributions in the context of the continuous-time domain and Laplace transform is included due to their importance in evaluating transforms.

Shift in s Plane

$$< G(s-s_0), \Phi(s) >_{\Re[s]=\sigma} = \int_{\sigma-j\infty}^{\sigma+j\infty} G(s-s_0)\Phi(s)ds$$

Letting $s - s_0 = y, \quad ds = dy$ we obtain

$$< G(s-s_0), \Phi(s) >_{\Re[s]=\sigma} \;=< G(y), \Phi(y+s_0) >_{\Re[y]=\sigma-\sigma_0}$$

Scaling

Let $\gamma \neq 0$ be a real constant. We can write

$$< G(\gamma s), \Phi(s) >_{\Re[s]=\sigma} = \int_{\sigma-j\infty}^{\sigma+j\infty} G(\gamma s)\Phi(s)ds$$

Letting $\gamma s = y, \quad \gamma ds = dy$ we obtain

$$< G(\gamma s), \Phi(s) >_{\Re[s]=\sigma} = \frac{1}{|\gamma|} < G(y), \Phi(y/\gamma) >_{\Re[y]=\gamma\sigma}$$

Product with an Ordinary Function

Consider the product $G(s)F(s)$. We can write

$$< G(s)F(s), \Phi(s) >_{\Re[s]=\sigma} \;=< G(s), F(s)\Phi(s) >_{\Re[s]=\sigma}$$

if

$F(s)\Phi(s) \in C$, the class of test functions.

Convolution

Denoting by $G_1(s) * G_2(s)$ the convolution of two generalized distributions, with $y = \Sigma + j\Omega$, we may write

$$I=<G_1(s)*G_2(s),\Phi(s)>_{\Re[s]=\sigma} \quad =<\int_{\Sigma-j\infty}^{\Sigma+j\infty} G_1(y)G_2(s-y)dy,\Phi(s)>_{\Re[s]=\sigma}$$

$$I=<G_1(y),\int_{\sigma-j\infty}^{\sigma+j\infty} G_2(s-y)\Phi(s)ds>_{\Re[y]=\Sigma}$$

the integral on the right, being in the form of a convolution with a test function, belongs to the class of test functions.

Derivation

$$<G'(s),\Phi(s)>_{\Re[s]=\sigma}=\int_{\sigma-j\infty}^{\sigma+j\infty} G'(s)\Phi(s)ds$$

Integrating by parts we obtain

$$<G'(s),\Phi(s)>_{\Re[s]=\sigma}=-<G(s),\Phi'(s)>_{\Re[s]=\sigma}$$

and, by repeated derivation,

$$<G^{(n)}(s),\Phi(s)>_{\Re[s]=\sigma} \quad =(-1)^n<G(s),\Phi^{(n)}(s)>_{\Re[s]=\sigma}$$

Multiplication of the Derivative Times an Ordinary Function

Consider the product $G'(s)F(s)$. We can write

$$<G'(s)F(s),\Phi(s)>_{\Re[s]=\sigma}$$
$$=\int_{\sigma-j\infty}^{\sigma+j\infty} G'(s)F(s)\Phi(s)ds$$

Integrating by parts we obtain

$$<G'(s)F(s),\Phi(s)>_{\Re[s]=\sigma}=-<G(s),F(s)\Phi'(s)>_{\Re[s]=\sigma}-<G(s),F'(s)\Phi(s)>_{\Re[s]=\sigma}$$

Generalized Distributions in Z Domain

A generalized distribution $G(z)$ in the z domain may be defined as the value of the integral, denoted $I_G[\Phi(z)]$, of its product with a test function $\Phi(z)$. Symbolically, we write

$$I_G[\Phi(z)]=<G(z),\Phi(z)>_{|z|=r} \quad = \oint_{|z|=r} G(z)\Phi(z)dz \tag{3}$$

where the contour of integration is a circle of radius r centred at the origin in the z plane. Similar properties to the continuous time domain are encountered in the discrete-time domain.

THE GENERALIZED DELTA IMPULSE IN THE S DOMAIN

The generalized Dirac-delta impulse denoted $\xi(s)$ may be defined by the relation

$$< \xi(s), \Phi(s) >_{\Re[s]=\sigma} = \begin{cases} \int_{-j\infty}^{j\infty} \xi(s)\Phi(s)ds = j\Phi(0), \sigma = 0 \\ 0, \quad ,\sigma \neq 0 \end{cases} \tag{5}$$

If $F(s)$ is analytic at $s=0$ then

$$< \xi(s), F(s) >_{\Re[s]=\sigma} = \begin{cases} \int_{-j\infty}^{j\infty} \xi(s)\Phi(s)ds = jF(0), \quad \sigma = 0 \\ 0, \quad ,\sigma \neq 0 \end{cases} \tag{6}$$

Some important properties are summarized in the following.

Derivation

$$< \xi^{(n)}(s-s_0), \Phi(s) >_{\Re[s]=\sigma} = \begin{cases} (-1)^n j\,\Phi^{(n)}(s_0), \quad \sigma = \sigma_0 \\ 0, \quad \sigma \neq \sigma_0 \end{cases} \tag{7}$$

Convolution

$$\xi(s-a) * \xi(s-b) = j\xi[s-(a+b)] \tag{8}$$

Convolution with an Ordinary Function

$$\xi(s-s_0) * F(s) = jF(s-s_0) \tag{9}$$

Multiplication of an Impulse Times an Ordinary Function

$$\xi(s-a)F(s) = F(a)\xi(s-a) \tag{10}$$

Multiplication by the nth Derivative of the Impulse

Applying the property of the derivative times an ordinary function we obtain

$$\xi'(s)F(s) = F(0)\xi'(s) - F'(0)\xi(s)$$

More generally we obtain

$$F(s)\xi^{(n)}(s) = \sum_{k=0}^{n}(-1)^k \binom{n}{k} F^{(k)}(0)\xi^{(n-k)}(s)$$

THE GENERALIZED IMPULSE IN *Z* DOMAIN

The discrete-time domain generalized impulse will be denoted by the symbol $\psi(z)$ and is equivalent to the symbol $\zeta(z-1)$ proposed earlier [16], that is,

$$\psi(z) = \zeta(z-1)$$

$$<\psi(z), \Phi(z)>_{|z|=r} = \begin{cases} j\Phi(1), r=1 \\ 0, r \neq 1 \end{cases}$$

If $X(z)$ is analytic at $z=1$ then

$$\oint_{|z|=r} \psi(z)F(z)dz = \begin{cases} jF(1), r=1 \\ 0, r \neq 1 \end{cases}$$

Derivation

$$<\psi^{(n)}(z),\ \Phi(z)>_{|z|=r} = \begin{cases} j(-1)^n \Phi^{(n)}(1), r=1 \\ 0, r \neq 1 \end{cases} \tag{11}$$

EXPONENTIAL IMPULSE TRAINS

A generalization of the periodic impulse train is an exponential impulse Trains. To evaluate its Laplace transform let

$$f(t) = e^{\alpha t}\rho_T(t) = e^{\alpha t}\sum_{n=-\infty}^{\infty}\delta(t-nT) = \sum_{n=-\infty}^{\infty} e^{\alpha nT}\,\delta(t-nT) \tag{12}$$

We have

$$F(s) = \frac{2\pi}{T}\sum_{n=-\infty}^{\infty} \xi\left(s - jn\frac{2\pi}{T} - \alpha\right) \tag{13}$$

We evaluate

$\mathcal{L}[x(t)]$, where $x(t) = \sum_{n=0}^{\infty} \delta(t - nT)$.

Let

$$v_s(t) = x(t) - 0.5\delta(t) = \sum_{n=0}^{\infty} \delta(t - nT) - 0.5\delta(t) \tag{14}$$

We may write [13]

$$v_s(t) = \rho_T(t)u(t) = \sum_{n=-\infty}^{\infty} \delta(t - nT)u(t) \tag{15}$$

$$V_s(s) = \sum_{n=0}^{\infty} e^{-nTs} - 0.5 = \frac{1}{1-e^{-Ts}} - 0.5, \quad \sigma > 0 = \frac{1}{2}\coth(Ts/2), \quad \sigma > 0$$

Moreover

$$V_s(s) = \frac{1}{2\pi}\left\{\sum_{n=-\infty}^{\infty} \frac{2\pi}{T}\xi\left(s - jn\frac{2\pi}{T}\right)\right\} * \left\{\frac{1}{s} + \pi\xi(s)\right\}$$

$$= \frac{1}{T}\left\{\sum_{n=-\infty}^{\infty} \frac{1}{s - jn\frac{2\pi}{T}} + \pi\xi\left(s - jn\frac{2\pi}{T}\right)\right\} \tag{16}$$

i.e.,

$$\rho_T(t)u(t) \leftrightarrow \frac{1}{T}\left\{\sum_{n=-\infty}^{\infty} \frac{1}{s - jn\frac{2\pi}{T}} + \pi\xi\left(s - jn\frac{2\pi}{T}\right)\right\} \tag{17}$$

Note that [13]

$$\frac{1}{T}\sum_{n=-\infty}^{\infty} \frac{1}{s + \alpha - j2\pi n/T} = \frac{1}{2}\coth\left[(s + \alpha)\frac{T}{2}\right] \tag{18}$$

i.e.,

$$\frac{1}{T}\sum_{n=-\infty}^{\infty} \frac{1}{s - j2\pi n/T} = \frac{1}{2}\coth(Ts/2) \tag{19}$$

Hence

$$V_s(s) = \frac{1}{2}\coth\left(\frac{Ts}{2}\right) + \frac{\pi}{T}\sum_{n=-\infty}^{\infty} \xi\left(s - jn\frac{2\pi}{T}\right) \tag{20}$$

$$= \frac{1}{1-e^{-Ts}} - 0.5 + \frac{\pi}{T}\sum_{n=-\infty}^{\infty} \xi\left(s - jn\frac{2\pi}{T}\right) \tag{21}$$

and

$$X(s) = V_s(s) + 0.5 \tag{22}$$

$$= \frac{1}{1-e^{-Ts}} + \frac{\pi}{T}\sum_{n=-\infty}^{\infty} \xi\left(s - jn\frac{2\pi}{T}\right) \tag{23}$$

Table 1. New Laplace Transforms

$x_c(t)$	**Extended $\xi-$ Transform $X_c(s)$**
1	$2\pi\xi(s)$
e^{at}	$2\pi\ \xi(s-a)$
$\cosh(at)$	$\pi\{\xi[s-a]+\xi[s+a]\}$
$\cosh(j\beta t)$	$\pi\{\delta[\omega-\beta]+\xi[\omega+\beta]\}$
$u(t)$	$1/s+\pi\xi(s)$
$e^{at}u(t)$	$1/(s-a)+\pi\xi(s-a)$
$e^{\alpha t}\cos(\beta t)$	$\pi\{\xi[s-(\alpha+j\beta)]+\xi[s-(\alpha-j\beta)]\}$
$e^{\alpha t}\cos\beta t\ u(t)$	$\dfrac{s-\alpha}{(s-\alpha)^2+\beta^2}+\dfrac{\pi}{2}\{\xi[s-(\alpha+j\beta)]+\xi[s-(\alpha-j\beta)]\}$
t	$-2\pi d\xi(s)/ds$
t^n	$(-1)^n 2\pi\xi^{(n)}(s)$
$t^n e^{at}u(t)$	$\dfrac{n!}{(s-a)^{n+1}}+(-1)^n\pi\xi^{(n)}(s-a)$
$1/(jt)+\pi\delta(t)$	$2\pi\nu(-s)$
$4\cos\beta t\cosh\alpha t$	$2\pi\{\xi(s-a)+\xi(s-a^*)+\xi(s+a)+\xi(s+a^*)\}$
$-tu(-t)$	$1/s^2+\pi\xi'(s)$
$t^n e^{\alpha t}u(t)$	$\dfrac{n!}{(s-\alpha)^{n+1}}+(-1)^n\pi\xi^{(n)}(s-\alpha)$
$(-t)^n e^{-\alpha t}u(-t)$	$\dfrac{n!}{(-1)^{n+1}(s+\alpha)^{n+1}}+\pi\xi^{(n)}(s+\alpha)$

$$\sum_{n=0}^{\infty} \delta(t-nT) \leftrightarrow \frac{1}{1-e^{-Ts}} + \frac{\pi}{T}\sum_{n=-\infty}^{\infty} \xi\left(s - jn\frac{2\pi}{T}\right). \tag{24}$$

and we have the alternative forms

$$\sum_{n=0}^{\infty} \delta(t-nT) \leftrightarrow 0.5 + \frac{1}{T}\sum_{n=-\infty}^{\infty} \frac{1}{s-\frac{j2\pi n}{T}} + \frac{\pi}{T}\sum_{n=-\infty}^{\infty} \xi\left(s - jn\frac{2\pi}{T}\right) \tag{25}$$

$$\sum_{n=0}^{\infty} \delta(t-nT) \leftrightarrow 0.5+\frac{1}{2}\coth(Ts/2)+\frac{\pi}{T}\sum_{n=-\infty}^{\infty} \xi\left(s-jn\frac{2\pi}{T}\right) \quad (26)$$

Table 2. New *z-Transforms*

$x[n]$	**Extended *z-Transform*** $X(z)$
1	$2\pi\psi(z)$
a^n	$2\pi\psi(z/a)$
$4\cos\gamma n\cosh\alpha n\{1+(-1)^n\}$	$2\pi\{\psi(z/a)+\psi(z/a^*)+\psi(z/a^{-1})+\psi(z/a^{-1*})$ $+\psi(-z/a)+\psi(-z/a^*)+\psi(-z/a^{-1})+\psi(-z/a^{-1*})\}$
$a^n\cos(\Omega_0 n)$	$\pi\left\{\psi\left[z/\left(ae^{j\Omega_0}\right)\right]+\psi\left[z/\left(ae^{-j\Omega_0}\right)\right]\right\}$
$u[n]$	$\frac{1}{1-z^{-1}}+\pi\psi(z)$
$a^n u[n]$	$\frac{1}{1-az^{-1}}+\pi\psi(z/a)$
$a^{n-n_0}u[n-n_0]$	$z^{-n_0}\left\{\frac{1}{1-az^{-1}}+\pi\psi(z/a)\right\}$
$a^n b^n u[n]$	$\frac{1}{1-abz^{-1}}+\pi\psi\{z/(ab)\}$
$\cos[\Omega_0 n+\phi]$	$\pi[e^{j\phi}\psi(z/e^{j\Omega_0})+e^{-j\phi}\psi(z/e^{-j\Omega_0})]$
$n^r u[n]$	$\sum_{i=0}^{r} S(r,i)\frac{i!}{(1-z^{-1})^{i+1}}z^{-i}$ $+(-1)^r\pi\sum_{i=0}^{r} S(r,i)z^i\psi^{(i)}(z^{-1})$
n^r	$(-1)^r 2\pi\sum_{i=0}^{r} S(r,i)\psi^{(i)}(z)$

Application to Impulse Invariance

In what follows we show that generalized distributions reveal al an unusual decades-old anomaly in the well-known digital filtering approach of impulse invariance. A description of the anomaly and the need for revisiting the sampling theorem was first noticed in [13] and has been recently presented using regular distributions [17, 18].

The Nyquist–Shannon sampling theorem, after the Swedish Electrical Engineering Scientist Harry Nyquist (1889-1976) and the American Mathematician Claude Elwood Shannon (1916-2001), is a fundamental result in the field of information theory, in particular telecommunications and signal processing. The theory is the basis for the transformation from continuous-time to discrete-time signals and systems. The presence of discontinuities in a sampled signal warrants a particular attention in applying the sampling theorem as will be presently shown. In fact, the theorem is based on the fact that given a continuous-time signal $x_c(t)$ which is ideally sampled by the impulse train $\rho_T(t)$, the result is the ideally sample signal

$$x_s(t) = x_c(t)\sum_{n=-\infty}^{\infty}\delta\,(t - nT) \tag{27}$$

and its Fourier spectrum is

$$X_s(j\omega) = \frac{1}{T}\sum_{n=-\infty}^{\infty} X_c\left[j(\omega - \frac{2n\pi}{T})\right] \tag{28}$$

This transform is then extended in the present literature to Laplace and z-domains. In what follows we focus our attention on signals containing discontinuities and show that Equation (28) does not necessarily hold true.

An important class of signals containing discontinuities is that of one-sided exponential signals. Causal exponential signals is one such case that warrants investigation. Similarly to the approach followed in Sec. 5 above we can evaluate the transforms of the sampled version of such signals. We shall focus our attention on the practical application of digital filter design using the well-known approach of Impulse Invariance. A an important consequence of the properties of periodic distributions as studied above is the revealing of an anomaly that presently exists in the literature in the formulation of Impulse Invariance. This approach is a major tool for the transformations of continuous-time to discrete-time systems. To illustrate the approach and the pitfall leading to the error consider a continuous-time system of transfer function

$$H_c(s) = \sum_{k=1}^{M}\frac{C_k}{s - p_k} \tag{29}$$

and let $C_k = A_k e^{j\theta_k}$. The impulse response is given by

$$h_c(t) = \sum_{k=1}^{M} C_k\, e^{p_k t} u(t) \tag{30}$$

Ideal sampling of the impulse response $h_c(t)$ with a sampling period T produces the signal $h_s(t)$ where

$$h_s(t) = h_c(t)\rho_T(t) \tag{31}$$

In the present day literature this expression is re-written in the form

$$h_s(t) = \sum_{k=1}^{M} C_k\, e^{p_k t}\sum_{n=0}^{\infty}\delta\,(t - nT) \tag{32}$$

and thus re-written as

$$h_s(t) = \sum_{k=1}^{M}\sum_{n=0}^{\infty} C_k\, e^{p_k nT}\delta(t - nT). \tag{33}$$

of which the Laplace transform is deduced as being

$$H_s(s) = \sum_{k=1}^{M} \sum_{n=0}^{\infty} C_k \, e^{p_k nT} e^{-nTs}. \tag{34}$$

i.e.,

$$H_s(s) = \sum_{k=1}^{M} \frac{C_k}{1-e^{-(s-p_k)T}}. \tag{35}$$

Moreover, from Equation (31), it is deduced that the Fourier transform of $h_s(t)$ is given by

$$H_s(j\omega) = \frac{1}{2\pi} H_c(j\omega) * \frac{2\pi}{T} \sum_{n=-\infty}^{\infty} \delta\left(t - \frac{n2\pi}{T}\right) = \frac{1}{T} \sum_{n=-\infty}^{\infty} H_c\left[j\left(\omega - \frac{n2\pi}{T}\right)\right] \tag{36}$$

It is subsequently stated that by replacing $j\omega$ by the more general Laplace variable s it follows that

$$H_s(s) = \frac{1}{T} \sum_{n=-\infty}^{\infty} H_c\left(s - j\frac{n2\pi}{T}\right) \tag{37}$$

Such transform extension from the $j\omega$ axis to the Laplace s domain is not justified since according to the current literature, and in the absence of the newly proposed generalized functions of a complex variables, the Laplace transform of the two-sided impulse train $\rho_T(t)$ simply does not exist. The Laplace transform of $h_s(t) = h_c(t)\rho_T(t)$ using convolution is therefore not mathematically justified. Such convolution can be effected only by employing the recently proposed generalized distributions.

The analysis is then extended in the literature to the discrete time domain where for impulse invariance the discrete domain impulse response is written

$$h[n] = h_c(nT) = \sum_{k=1}^{M} C_k \, e^{p_k nT} u[n]. \tag{38}$$

and its z transform, the system transfer function

$$H(z) = \sum_{k=1}^{M} \frac{C_k}{1-e^{p_k T} z^{-1}} \tag{39}$$

and frequency response

$$H(e^{j\Omega}) = \sum_{k=1}^{M} \frac{C_k}{1-e^{p_k T} e^{-j\Omega}}. \tag{40}$$

With $z = e^{Ts}$ it follows

$$H(e^{Ts}) = \sum_{k=1}^{M} \frac{C_k}{1-e^{p_k T} z^{-1}} = H_s(s). \tag{41}$$

so that

$$H(e^{Ts}) = \frac{1}{T} \sum_{n=-\infty}^{\infty} H_c\left(s - j\frac{n2\pi}{T}\right). \tag{42}$$

and with $z = e^{j\Omega} = e^{j\omega T}$, it follows that

$$H(e^{j\Omega}) = \frac{1}{T}\sum_{n=-\infty}^{\infty} H_c\left[j\frac{\Omega - 2n\pi}{T}\right]. \tag{43}$$

Comparing these results we note that

$$H(e^{Ts}) = H_s(s) = \sum_{k=1}^{M} \frac{C_k}{1-e^{-(s-p_k)T}} = \sum_{k=1}^{M} \frac{1}{T}\sum_{n=-\infty}^{\infty} \frac{C_k}{s-p_k-j2n\pi/T} \tag{44}$$

that is

$$\frac{1}{T}\sum_{n=-\infty}^{\infty} \frac{1}{s-p_k-j2n\pi/T} = \frac{1}{1-e^{-(s-p_k)T}} \tag{45}$$

The Mittag-Leffler's Expansion, after the Swedish mathematician Magnus Gösta Mittag-Leffler (Stockholm 1846-1927), contradicts this statement. In fact the Mittag-Leffler expansion states that, with z a generally complex variable,

$$\frac{1}{z} + 2z\sum_{n=1}^{\infty} \frac{1}{z^2+n^2\pi^2} = \coth\, z. \tag{46}$$

The expansion can be re-written in the form

$$\frac{1}{T}\sum_{n=-\infty}^{\infty} \frac{1}{z-j2n\pi/T} = \frac{1}{2}\coth\left(\frac{Tz}{2}\right) \tag{47}$$

which is not the value given in Equation (45). This reveals an error that has remarkably lain undetected for decades.

A possible explanation for the existence of such an old anomaly is in the fact that the multiplication of distributions is not well defined [13]. Such multiplication is implied in Equation (31) and leads to such mathematical inconsistency that is revealed by the Mittag-Leffler expansion. The proper approach to impulse invariance, in fact the proper approach to sampling one-sided functions, is to proceed as in Sec. 8 above, using the Mittag-Leffler expansion in order to resolve the ambiguity arising from the multiplication of distributions. We may write

$$h_s(t) = \sum_{k=1}^{M} C_k\, e^{p_k t} u(t)\rho_T(t) \tag{48}$$

and referring to Equation (17) we may write

$$H_s(s) = \sum_{k=1}^{M} C_k \left[\frac{1}{T}\left\{\sum_{n=-\infty}^{\infty} \frac{1}{s-p_k-jn\frac{2\pi}{T}} + \pi\xi\left(s - p_k - jn\frac{2\pi}{T}\right)\right\}\right] \tag{49}$$

Moreover,

$$H_s(s) = \sum_{k=1}^{M} C_k \left[\frac{1}{2}\coth\left[T(s-p_k)/2\right] + \frac{\pi}{T}\sum_{n=-\infty}^{\infty} \xi\left(s - p_k - jn\frac{2\pi}{T}\right)\right] \tag{50}$$

$$H_s(s) = \sum_{k=1}^{M} \frac{C_k}{1-e^{-(s-p_k)T}} - 0.5C_k + \frac{\pi}{T} C_k \sum_{n=-\infty}^{\infty} \xi \left(s - p_k - jn\frac{2\pi}{T}\right) \tag{51}$$

$$H_s(j\omega) = \sum_{k=1}^{M} \frac{C_k}{2} \coth\left[(j\omega - p_k)\frac{T}{2}\right], \Re[p_k] < 0 \tag{52}$$

and

$$h_s(t) = \sum_{k=1}^{M} C_k \, e^{p_k t} \sum_{n=0}^{\infty} \delta\,(t - nT) - 0.5C_k\delta(t) \tag{53}$$

$$h[n] = h_c(nT) = \sum_{k=1}^{M} C_k \left[e^{p_k nT} u[n] - 0.5\delta[n]\right]. \tag{54}$$

and its z transform, the system transfer function

$$H(z) = \sum_{k=1}^{M} \frac{C_k}{2} \frac{1+e^{p_k T} z^{-1}}{1-e^{p_k T} z^{-1}}. \tag{55}$$

and the frequency response

$$H(e^{j\Omega}) = \sum_{k=1}^{M} \frac{C_k}{2} \frac{1+e^{p_k T} e^{-j\Omega}}{1-e^{p_k T} e^{-j\Omega}} \tag{56}$$

Equations (55) and (56) are the true values of the digital filter transfer function and its frequency response obtained by a true application of impulse invariance and should replace the corresponding expressions Equations (39) and (40) found in the present day literature.

Conclusion

Important properties of the proposed generalized distributions have been extended and applied to the sampling of signals containing discontinuities. Applications ranging from evaluating transforms of functions and sequences that have hitherto had no transform to new formalism related to transforms of sampled signals. The important digital signal processing approach to continuous-time to discrete-time signal and system transformation lead to rewriting the transforms of functions containing discontinuities. The generalized distributions are shown to reveal an unusual error that escaped detection for decades in the digital filter design approach of impulse invariance. The Mittag-Leffler Expansion is invoked, revealing the existence of an anomaly leading to a basic mathematical error. It is subsequently used to prove the correction that needs be applied to eliminate the anomaly. The large deviation error produced by the common approach is evaluated in the context of a first order filter prototype and a general order filter. The vast improvement in performance of the proposed filter structures can be seen in [17].

ACKNOWLEDGMENT

Special Thanks are due to the National Science and Engineering Research Council NSERC of Canada for supporting this research.

REFERENCES

[1] Guest P. B., *"Laplace Transforms and Introduction to Distributions,"* Ellis Horwood, New York, 1991.
[2] Davies B., *"Integral Transforms and their Applications,"* Springer Verlag, New York, 1985.
[3] Donoghue, William F. Jr., *"Distributions and Fourier Transforms,"* Academic Press, New York, 1969.
[4] Barros-Neto José, *"An Introduction to the Theory of Distributions,"* Marcel Dekker, Inc., New York, 1973.
[5] Friedlander F. G. and Joshi M., *"Introduction to the Theory of Distributions,"* Cambridge University Press, 1998.
[6] Cristescu Romulus and Marinescu Gheorghie, *"Applications of the Theory of Distributions,"* Editura Academiei, John Wiley, London, 1973.
[7] Bracewell Ronald N., *"The Fourier Transform and Its Applications,"* McGraw Hill, New York, 1978.
[8] Schwartz Laurent, *"Méthodes mathématiques,"* Hermann, Paris, 1979.
[9] Tzannes Nicolaos S., *"Communication and Radar Systems,"* Prentice Hall, Englewood Cliffs, N.J., 1985.
[10] Poularikas Alexander D. and Seely Samuel, *"Signals and Systems,"* PWS-KENT, Boston, 1991.
[11] Papoulis Athanasios, *"The Fourier Integral and Its Applications,"* McGraw Hill, New York, 1962.
[12] Poularikas Alexander, Editor-in-Chief, *"The Transforms and Applications Handbook,"* CRC Press, IEEE Press, Boca Raton, 2000.
[13] Corinthios M., *"Signals, Systems, Transforms and Digital Signal Processing with MATLAB©,"* Taylor and Francis (CRC), June 2009.
[14] Corinthios M., *"New Laplace, z and Fourier-Related Transforms,"* Proceedings of the Royal Society A: Mathematical, Physical and Engineering Sciences, UK, Proc. Roy. Soc. A, (2007) vol. 463 pp. 1179-1198, May. 2007.
[15] Corinthios M., *"Complex-variable Distribution theory for Laplace and z-Transforms,"* IEE Proc. Vision, Image and Signal Processing, Vol. 152, N0. 1, Feb. 2005, pp. 97-106.
[16] Corinthios M., *"Generalisation of the Dirac-delta impulse extending Laplace and z-Transform domains,"* IEE Proc. Vision, Image and Signal Processing, Vol. 150 No. 2, April 2003, pp. 69-81.

[17] Corinthios M., *"Application of the Mittag–Leffler expansion to sampling discontinuous signals,"* IET Signal Processing, Dec. 2013, Vol. 7, Iss. 9, pp. 863-878, Dec. 2013.

[18] Corinthios M., *"Generalized distributions, sampling theorem revisited and an end to an impulse invariance error,"* IET 2nd Intl. Conf. on Intelligent Signal Process, Kensington Close Hotel, London, U.K. 1-2 Dec. 2015.

In: Digital Signal Processing (DSP)
Editor: Juan Zhang

ISBN: 978-1-63485-168-8

Chapter 2

ACTIVE CONTOUR MODELS FOR IMAGE SEGMENTATION OF COMPLEX OBJECT SHAPES

Shuqun Zhang*

Department of Computer Science, College of Staten Island,
City University of New York, US

ABSTRACT

Active contour models have been widely applied in a variety of image processing and computer vision tasks such as image segmentation and object tracking. However they usually have poor performance in segmenting images with complex object shapes and handling weak-edge-leakage. This chapter summarizes some recent techniques for effectively driving active contours towards complex object boundaries and overcoming the weak-edge-leakage problem. These techniques incorporate useful information such as evolution direction prior, object centerline, geometric information of edge map, saddle and stationary point detection result into active contour model design, and produce novel external force fields for improved performance. The techniques for segmenting complex shapes are mainly based on providing additional force towards the desired evolution direction, and the methods for weak-edge-leakage are mainly based on changing the field forces near the edge to point perpendicularly to the weak edge. Various image segmentation experimental results are given to show the good performance of these active contour models.

Keywords: image segmentation, active contours

INTRODUCTION

Image segmentation is to partition an image into several regions of interest such that each region has similar features of gray-level or texture. It is one of the most important yet difficult problems in image processing and computer vision, and thus has been widely studied. In

---

* Email: shuqun.zhang@csi.cuny.edu.

particular, accurate image segmentation plays a crucial role in biomedical imaging applications, which makes rendering of 3D biological structure and automatic quantitative analysis of regions of interest possible. Although numerous image segmentation techniques have been proposed, it is still a complex and challenging problem for many applications. This could be due to image quality, intensity inhomogeneity, presence of noise/clutter/occlusion, object complexity and other factors. Image segmentation techniques may include threshold-based segmentation, edge-based segmentation, region-based segmentation, clustering based segmentation, recognition-based segmentation, partial differential equation (PDF)-based segmentation, neural network-based segmentation, and graph-based segmentation. In this chapter, we focus on one of the PDE framework-based methods, active contour model, which was first introduced by Kass et al. [1] and has been widely used for biomedical image segmentation.

Active contour, also known as snake, is a dynamic curve modeled to evolve towards the object boundary by minimizing some given energy functional. The evolution of an active contour is driven by the internal force set to regularize the dynamic curve and the external force set to attract the dynamic curve to the desired image features, such as the image edge formed at the object boundary. If the curve is properly initialized and the parameters are well tuned, the initial contour will be evolved toward the expected object boundary in the absence of strong noise and clutter. When any one of the above conditions fails, the contour may have problem in converging toward the object boundary. Therefore many modifications of the original snake have been proposed to overcome these problems under different constraints. Since the external force is critical for the performance of an active contour, most research in the field has focused on the design of external force or field, for example Cohen and Cohen's balloon force [2], Park and Chung's virtual electric field (VEF) [3], Xie and Mirmehdi's magnetostatic field [4], Sum and Cheung's boundary vector field [5], Li and Acton's vector field convolution (VFC) [6], Xu and Prince's gradient vector flow (GVF) field [7] and its variants such as the generalized GVF [8], the multi-scale GVF field [9], the extended GVF field [10] for downstream algorithm, and region-aided GVF[11].

In designing external forces, various priors such as object shape, size, position, motion, and/or direction can be used to incorporate into active contours to further improve the performance of active contour models. For example, Ray et al. [12] incorporated the motion direction into the GVF model for tracking rolling leukocytes. The GVF field is also modified by considering the directional information of image edge to make active contours discern the image edge with different directions [13, 14]. Considering that prior information is not always available for all images, and maybe difficult to obtain or estimate, a more general method called gradient and direction vector flow (G&DVF) [15] for providing prior direction information was recently proposed, which is implemented by drawing a few lines towards the desired evolving direction after the normal contour initialization step, which can significantly improve the performance of the GVF model on segmenting images with complex object shape, and can alleviate the requirement that the initial contour must be very close to the true object boundary in order to obtain good performance.

Performance of active contour models generally depends on proper initialization and selection of forces and energy functions. The main problems of active contour-based image segmentation have been the poor performance in segmenting images with complex object shape and dealing with weak-edge-leakage. This chapter first briefly reviews the traditional active contour model and the GVF field, then explains the two above-mentioned problems, and finally presents some recent techniques for overcoming these two drawbacks. These solutions extend the capability of the GVF and VEP fields, and combine it with other force fields for improved performance in segmenting images with complex shape and weak edge.

BACKGROUND

The traditional active contour [1] is modeled to evolve a dynamic curve $\mathbf{x}(s) = [x(s), y(s)]$, $s \in [0,1]$ from an appropriate initial position to the object boundary by minimizing the following energy functional

$$E_{\mathrm{AC}}(\mathbf{x}) = \frac{1}{2}\int_0^1 \alpha|\mathbf{x}'(s)|^2 + \beta|\mathbf{x}''(s)|^2 \, ds + \int_0^1 E_{\mathrm{Ext}}(\mathbf{x}(s))\, ds, \tag{1}$$

where α and β are positive weighting parameters, and the function E_{Ext} is the external energy. The first term is referred to as the internal energy, which controls the smoothness of the curve $\mathbf{x}$, while the second term is referred to as the external energy, which attracts the curve $\mathbf{x}$ toward the object boundary. The traditional active contour uses the negative intensity of the image edge map f as the external energy, which is computed by first smoothing the image I with a Gaussian kernel followed by a gradient to enhance the boundaries as

$$E_{\mathrm{Ext}}(x,y) = -f(x,y) = -\left|\nabla\left[G_\sigma(x,y) * I(x,y)\right]\right|^2, \tag{2}$$

where $G_\sigma(x, y)$ denotes a Gaussian filter with standard deviation σ. The curve is evolved by minimizing the functional E_{AC} with respect to $\mathbf{x}$, which is considered as a function of s and artificial time t, and is implemented via the gradient descent

$$\mathbf{x}_t(s,t) = \left[\alpha\mathbf{x}''(s,t) - \beta\mathbf{x}''''(s,t)\right] - \nabla E_{\mathrm{Ext}}, \tag{3}$$

where the first and the second terms are referred to as the internal force and the external force, respectively.

The GVF [7] is the most widely investigated external force field for active contours. It was proposed to substitute the traditional external force in Eq. (3) to have larger capture range over the traditional active contour. Let $\mathbf{v}(x,y) = [u(x,y), v(x,y)]$ denote the GVF field, then it is computed as the diffusion of the gradient vectors of the image edge map by minimizing the following energy functional

$$E_{\mathrm{GVF}}(\mathbf{v}) = \iint \mu|\nabla\mathbf{v}|^2 + |\nabla f|^2 |\mathbf{v} - \nabla f|^2 \, dxdy. \tag{4}$$

The first term in the integrand is used to smooth the vector field $\mathbf{v}$, and has the effects of making the force field robust to image noises and enlarging the capture range of the force field. μ is the smoothness regularization parameter. Whereas, the second term is the data fidelity term that keeps **v** being equal to the gradient vector of the edge map ∇f. Similar to the minimization of E_{AC}, the GVF field can be obtained by solving the following Euler-Lagrange equation

$$\mathbf{v}_t = \mu\nabla^2\mathbf{v}(t) - (\mathbf{v} - \nabla f)|\nabla f|^2, \tag{5}$$

which is derived from the variational minimization of the energy functional E_{GVF} with respect to **v**.

Besides larger capture range, the GVF also has the ability of pushing the fronts of active contours into simple concave boundaries. These two advantages are very important because it can free the user from the burden of manual contour initialization and improve the segmentation performance. However, the GVF active contour still has the problem in converging to the object boundary with more complex shape.

There exist many different complex object shapes in natural images. Figure 1 shows some typical complex shapes. Since parametric active contours can't automatically handle the topology change of contour, here the object shape for active contours only refers to a shape represented by one closed curve, and it does not include the shape represented by more than one closed curve such as a ring shape (needs two). Generally, the shapes for active contours can be classified into two types: convexity and concave. It is well known that the conventional active contours can easily conform to a convexity, but they have difficulties in conforming to concaves except the 'V' and 'U' shaped concaves. The following figure shows a variety of concaves that most active contours cannot handle well.

Several efforts have been made to active contours to handle the segmentation of images with complex object shape. The simplest method is through manual contour initialization by placing an initial contour very close to the true object boundary, which is a difficult task. To relieve the initialization of active contours, automatic method is desired by designing novel external forces. For example, the balloon force [2] has the nice feature of enforcing an active contour to expand (or shrink) towards the object boundary even with complex shape by analogizing the inflation of a balloon. However, it can only evolve the active contour in one direction (i.e, either expanding or shrinking). Compared with the balloon force, the GVF can drive active contour to bidirectionally evolve towards the object boundary. However the GVF active contour model can only conform to the 'U' shaped concave, and it has difficulty in conforming to the semi-closed, 'S' shaped, 'L' shaped, screwy and hooked concaves. The NGVF [16] improves the convergence into long, thin indention and the convergence speed of the GVF with larger time step, but still cannot handle various complex shapes.

Another major drawback of active contours is the leakage due to weak edges. For example, balloon force can drive an active contour into complex shapes but it can also easily cause the contour to leak through a weak boundary. Region-based active contours [11, 17] are more success in overcoming the boundary leakage problem than edge-based active contours. There are not many solutions for the weak-edge-leakage problem for edge-based active contours. Mixing the VFC field [6] with the standard external force can alleviate the problem a little bit but noise performance will be affected. This chapter describes two recent methods for edge-based active contours.

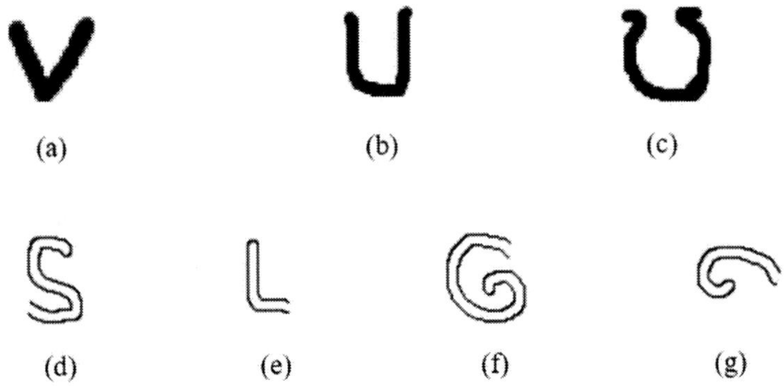

Figure 1. Various concave shapes. (a) 'V' shaped concave, (b) 'U' shaped concave, (c) semi-closed concave (i.e., 'C' shaped concave), (d) 'S' shaped concave, (e) 'L' shaped concave, (f) Screwy concave, and (g) Hooked concave.

METHODS

Many energy functions and forces for active contours cannot effectively deal with a variety of complex object shapes that appears in natural images. To find out the reason, we can exam the distribution of GVF field. A simple example is given in Figure 2, where we can see that for a complex shape there normally exit saddle points and/or stationary points in its GVF field.

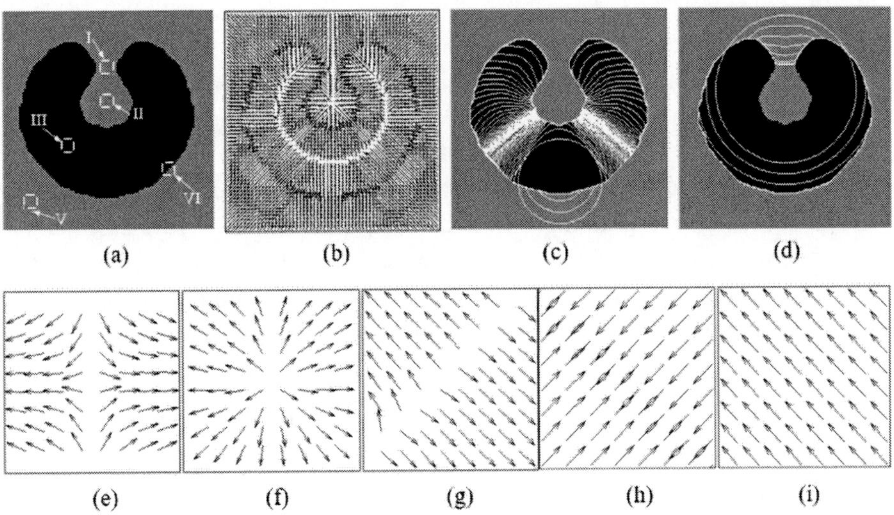

Figure 2. GVF field. (a) Original synthetic image. (b) GVF field. (c) and (d) Evolutions of the GVF active contours from two different initializations. (e)-(i) Closed-ups of the GVF field at the points 'I' to 'V' marked in (a).

Figures 2(a) and 2(b) show the image to be segmented and its GVF, respectively. The shape of the black object is like a supine 'c' with semi-closed concave. Figures 2(e) and 2(g) show the closed-ups of the saddle points formed outside and inside the object boundary, respectively. Figure 2(f) is the closed-up of the stationary point formed at the center of the semi-closed concave. The closed-ups of the GVF field at the image edge and the smooth areas are given in Figures 2(h) and 2(i), respectively. Figures 2(c) and 2(d) show the evolution of the GVF active contour with two different initializations, respectively. It is seen that when an active contour is set to evolve onto the saddle points, it might be stuck there as shown in Figures 2(c). And when there exit stationary points between the active contour and the object boundary, it is also impossible for the active contour to move across the stationary points as shown in Figure 2(d). In both situations the GVF snake has trouble in evolving towards the object boundary.

GVF-Based Balloon Force

To overcome the above-identified problem, obviously an additional force is needed to continuously push the active contour forward when it is stuck at saddle and/or stationary points. The balloon force is good candidate for this additional force because it can enforce active contour to expand (or shrink) in the absence of external influences towards the object boundary, no matter how complex the shape of the object is. However, there is a problem with the balloon force. It must be be initialized completely inside the object for the outward evolution or completely outside the object for the inward evolution. This unidirectional evolution brings the inconvenience to the initialization. For example in the application of object tracking using active contours, the final contour derived from the previous frame is generally used as the initial contour for the current frame, which is very likely across the object boundary. In this case the active contour using the balloon force will perform poorly. On the other hand, the GVF field can drive active contours to bidirectionally evolve towards the object boundary. It is therefore very nature to combine the advantages of the GVF field and the balloon force, and to overcome their respective problems. In this bidirectional evolution GVF-based balloon force [18], the GVF is used to move acive contour except when the curve meets saddle and stationary points where the balloon force is applied.

Obviously a saddle and stationary point detector is needed to develop for the GVF-based ballon force. It is seen from the example GVF distribution shown in Figure 2(b) that the GVF field is smooth in most image domain, except at the saddle and stationary points and along the strong image edge. It is also noted that the magnitude of the gradient of GVF field, i.e., $|\nabla \mathbf{v}(x, y)|$, will be approximated to zero where the GVF field is smooth, and in other places it will have a relatively large value. Therefore, one can discern the saddle and stationary points as well as the edge points from other smooth GVF field according to the value of $|\nabla \mathbf{v}(x, y)|$. To further separate the saddle and stationary points from edge points, an edge indicator function $g(x, y) = 1/[1 + f(x, y)]$ is used, which is very small when the pixel at (x, y) is on the strong edge and has a value of one in other places. Therefore, a weighting function map can be computed that can sepearte the saddle and stationary points in the GVF field from other (smooth GVF field and image edge) pixels in the image as

$$k(x, y) = g(x, y)|\nabla \mathbf{v}(x, y)| = g\sqrt{u_x^2 + u_y^2 + v_x^2 + v_y^2}. \tag{6}$$

The weighting function map for Figure 2(a) is obtained as shown in Figure 3, which clearly shows the saddle and stationary points. The bidirectional evolution balloon force derived from the GVF field is thus can be defined by

$$F_{\mathrm{B}}(\mathbf{x}) = \eta k(\mathbf{x})\bar{N}(\mathbf{x}), \tag{7}$$

where $\bar{N}(\mathbf{x})$ is the outward unit normal vector along the active contour $\mathbf{x}(s)$, and η is a pre-set constant parameter, which takes a positive value for expanding active contours, and a negative value for shrinking active contours. The balloon force can enforce active contours to expand (or shrink) towards the object boundary only when the active contours meet the saddle and/or stationary points in the GVF field. It thus can maintain the bidirectional evolution property of the GVF field, and meanwhile assist the GVF field in driving active contours to evolve towards the object boundary with certain complex shape. By replacing the external force $-E_{\mathrm{Ext}}$ in the traditional parametric active contour model with the combined force, a modified active contour can be obtained as

$$\mathbf{x}_t(s,t) = \left[\alpha\mathbf{x}''(s,t) - \beta\mathbf{x}''''(s,t)\right] + \mathbf{v}(\mathbf{x}) + F_{\mathrm{B}}(\mathbf{x}). \tag{8}$$

The segmentation results of the ‘C’ shape image from two different initializations are provided as shown in Figures 3(a) and 3(b), respectively. It is seen that the active contours with the GVF-based balloom force can quickly conform to the exact boundary.

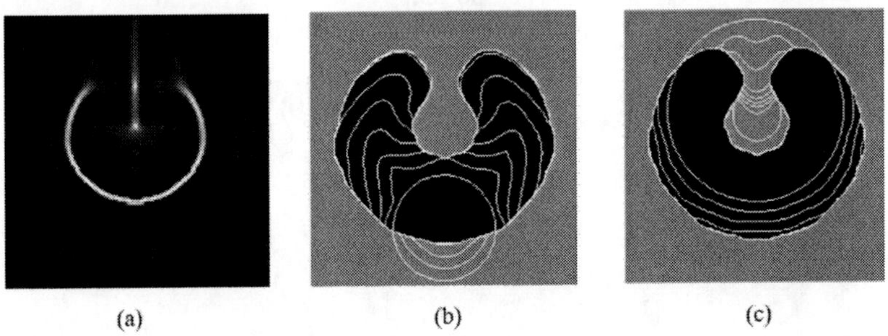

Figure 3. (a) Weighting map of a bidirectional balloon force for the image shown in Figure 2(a). (b) and (c) Evolutions of the GVF-based balloon force active contours from two different initializations.

More resutls are provided below to further show the good performance of the active contours using the bidirectional GVF-based balloon force in segmenting complex shapes.

Figure 4 shows the segmentation results of the GVF-based balloon active contour on three different synthetic complex shape images, which are also compared with the results obtained with the original GVF active contour and the adaptive balloon active contour [19]. For each of the three tested images, the initializations of the three active contours are the same (the small circle). The evolutions of the GVF active contour, the adaptive balloon active contour and the bidirectional GVF-based active contour with expanding balloon force are given in the first, second and third rows, respectively. It is seen that neither the GVF active contour nor the adaptive balloon active contour can correctly extract the three object boundaries with complex shapes, while the bidirectional balloon active contour can converge to the object boundaries successfully. Segmentation results from some real images given in Figure 5 also support the better performance of the GVF-based balloon active contour over the GVF active contour and the adaptive balloon active contour.

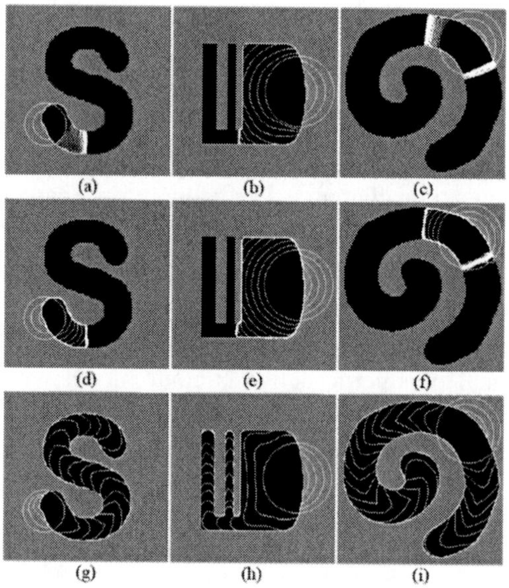

Figure 4. Segmentation results of three complex shapes by the GVF active contour, the adaptive balloon active contour and the bidirectional GVF-based balloon active contour, respectively. (a)-(c) Evolutions of the GVF active contour. (d)-(f) Evolutions of the adaptive balloon active contour. (g)-(i) Evolutions of the bidirectional GVF-based balloon active contour with expanding balloon force.

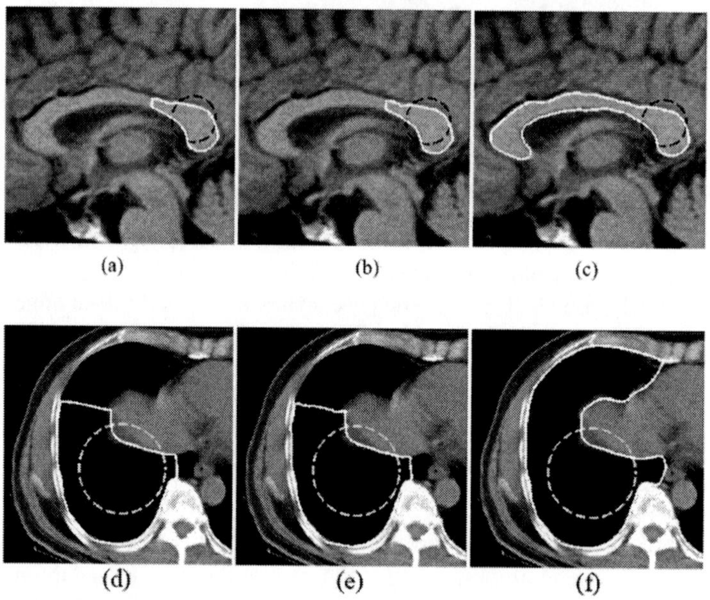

Figure 5. Segmentation results of two real images by the GVF active contour ((a) and (d)), the adaptive balloon active contour ((b) and (e)), and the bidirectional GVF-based balloon active contour ((c) and (f)), respectively.

Gradient and Direction Vector Flow

The G&DVF [15] is another recently proposed active contour that can effectively segment complex shapes. It uses a simple method to guide the contour toward the correct object boundary by providing the desired evolving direction to the contour as a prior, which can be simply implemented by drawing a few directional lines after the normal contour initialization step. The G&DVF seamlessly integrates the directional vector field produced by the directional lines with the GVF field, and improves the effectiveness of the GVF active contour model in segmenting complex shapes. The direction vector field dominates the GVF field at the stationary and saddle points, and can easily drive the active contour forward according to the preset direction.

The G&DVF is defined as the vector field $\mathbf{v}(x,y) = [u(x,y), v(x,y)]$ that minimizes the following energy functional

$$E_{\mathrm{G\&DVF}}(\mathbf{v}) = E_{\mathrm{GVF}}(\mathbf{v}) + \eta E_{\mathrm{DVF}}(\mathbf{v}) \tag{9}$$

where E_{GVF} and E_{DVF} denote the GVF functional and the direction vector flow (DVF) functional that is generated by the directional lines, respectively, and η is a positive weighting parameter that controls the influence of the directional lines on the external force field $\mathbf{v}$. The functional E_{DVF} is computed as

$$E_{\mathrm{DVF}}(\mathbf{v}) = \iint |\mathbf{w}|^2 \, |\mathbf{v} - \mathbf{w}|^2 \, dxdy, \tag{10}$$

where $\mathbf{w}$ is the DVF field obtained by two user mouse clicks in the image, and is represented as

$$\mathbf{w}(x,y) = \begin{cases} \dfrac{\mathbf{b}-\mathbf{a}}{|\mathbf{b}-\mathbf{a}|}, & \text{if } (x,y) \text{ is on } \overrightarrow{\mathbf{ab}} \\ \mathbf{0}, & \text{otherwise,} \end{cases} \tag{11}$$

where $\mathbf{a}$ and $\mathbf{b}$ denote the starting and the ending point of the directional line $\overrightarrow{\mathbf{ab}}$, respectively. If multiple directional lines are drawn, the DVF field is simply obtained by adding all the individual direction vector fields together as

$$\mathbf{w} = \mathbf{w}_1 + \cdots + \mathbf{w}_N, \tag{12}$$

where N depends on the complexity of the image and object shape. Normally one or two will be enough in most cases.

The DVF functional in Eq. (10) is designed to keep $\mathbf{v}$ being equal to $\mathbf{w}$ where the norm of $\mathbf{w}$ is relatively large. The G&DVF field is obtained by solving the following Euler-Lagrange equation

$$\mu \nabla^2 \mathbf{v} - (\mathbf{v} - \nabla f)|\nabla f|^2 - \eta(\mathbf{v} - \mathbf{w})|\mathbf{w}|^2 = 0. \tag{13}$$

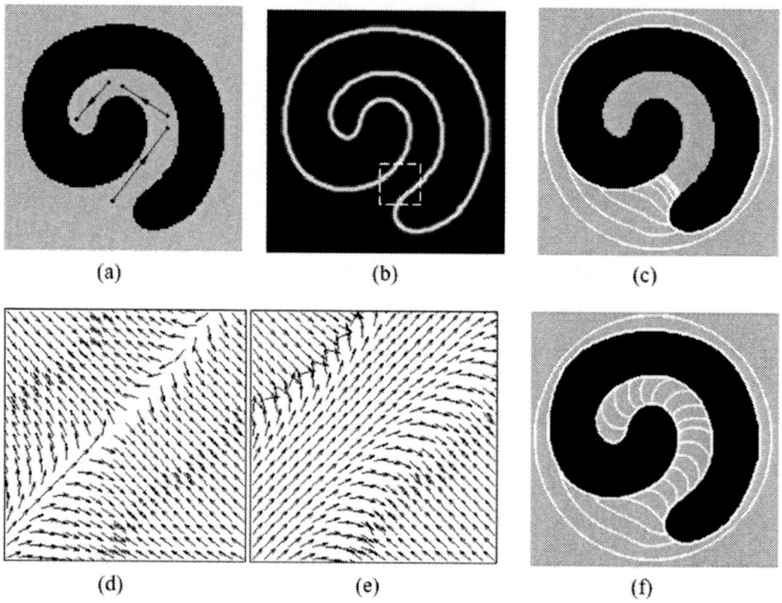

Figure 6. Segmentation of a swirl shaped image. (a) Original image, on which three directional lines are drawn for the G&DVF. (b) Edge map of the image. (c) Evolution of the GVF active contour. (d) and (e) the GVF field and the G&DVF field within the dashed area of (b), respectively. (f) Evolution of the G&DVF active contour.

The G&DVF active contour has been demonstrated to be very easy for operation, and significantly improve the performance of the GVF model on segmenting images with complex object shape as well as images containing clutters [15]. Figure 6 shows the results of the G&DVF and the original GVF in segmenting a complex swirl shape. It is seen from Figure 6(c) that the GVF active contour fails to conform to the object boundary from the big circle initialization. For the G&DVF active contour, inside the rolling concave three directional lines are added, as shown in Figure 6(a). These three short lines make almost all field force vectors that point to different directions (Figure 6(d)) inside the rolling concave point inward as shown in Figure 6(e). This field force change effectively drives the front of the G&DVF active contour to move into the concave as shown in Figure 6(f) to achieve perfect segmentation.

The G&DVF field is universal for any type of images and very easy to implement because it is simply implemented by drawing lines on the image to be segmented. It can help active contour go into any complex shape, and alleviate the requirement that the initial contour must be close to the true object boundary. Figure 7 shows more comparing results between the GVF and G&DVF active contours, which are performed on a magnetic resonance (MR) image of corpus-callosum and a femur CT image. The active contours are all initialized with a circle. The corpus-callosum image shown in Figure 7(a) contains a hooked concave, in which the GVF field has the problem in driving the curve since the front of the GVF active contour stops moving at the beginning of the concave, as shown in Figure 7(b). After drawing three directional lines for the G&DVF field, it is seen from Figure 7(c) that the G&DVF field successfully drives the fronts of the active contour to conform to the object boundary. For segmenting the femur CT image given in Figure 7(d), the front of the GVF active contour is attracted by the wrong boundary and is prevented from expanding to the femur boundary by the image edges formed by the clutters as shown in Figure 7(e). Again after placing two directional lines on the image for the G&DVF field, the G&DVF active contour converges to the outside boundary of the femur as expected. This is because the local distribution of the field around the clutters is now dominated by the DVF field formed by the two directional lines.

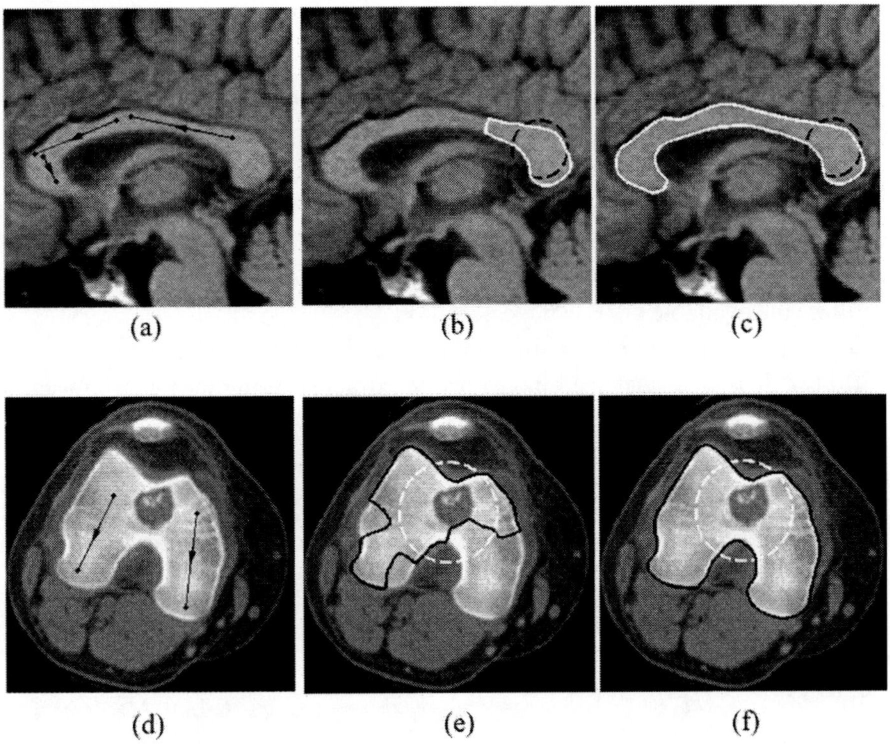

Figure 7. Segmentation results of a corpus-callosum MR image and a femur CT image: (a) and (d) Original image, on which directional lines are used for the G&DVF. (b) and (e) Segmentation of the GVF active contour. (c) and (f) Segmentation of the G&DVF active contour.

In the G&DVF active contour, although prior evolution direction information can be applied to all images because it is provided manually by mouse clicks, a problem with this method is that users must know where on the image to draw the directional lines, which is sometimes inconvenient. Therefore it is desired to develop an automatic method to extract the same or similar prior information from the image as the manual method provides. Developing such an automatic method for prior direction extraction is important for the G&DVF active contours because it can eliminate extra user input and avoid contour deviation due to inexperienced user's input. In fact it is not difficult to see that the G&DVF model will have the best segmentation performance if the directional lines are drawn along the centerline of a concave or object. It was therefore proposed to extract the centerlines automatically to help users in using G&DVF [20], which can use the same method for detecting saddle/stationary points described above.

The weighting function map obtained by detecting saddle/stationary points actually can be directly used to extract the centerline of an object or concave. This can be seen from Figure 3(a), where the weighting function map for the balloon force contains the centerlines for the object and concave. Therefore, we can extract the centerlines by first normalizing the weighting function map to [0, 1] and then comparing with a threshold

$$s(x,y)=\begin{cases}1 & if\left(\dfrac{k(x,y)-\min(k)}{\max(k)-\min(k)}\right)^{q}>T,\\ 0 & else,\end{cases} \tag{14}$$

where T is a pre-set threshold between 0 and 1, and q is the field strength taking a value between 0 and 1, which can be used to control the centerline strength. Morphological operations can be further applied to process the thresholded result, and skeleton algorithm can be used to make lines thinner. Once the centerlines are extracted, the G&DVF users will know where the best place is to draw the directional lines and generate the corresponding DVF field. This centerline detection method has very low computational complexity since it directly utilizes the available GVF in the calculation.

Figures 8(d)-(f) show the centerline extraction results for the three different images shown in Figures 8(a)-(c). It is seen that the automatically extracted centerlines can be used to help draw directional lines for the G&DVF (shown as gray arrow lines), which is very similar to those manually drawn directional lines in Figures 8(a)-(c).

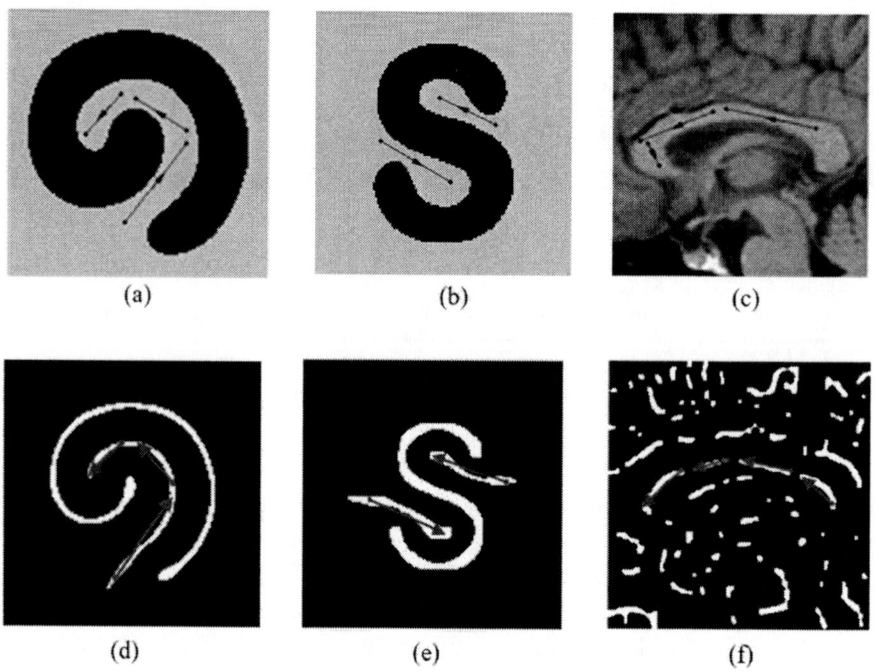

Figure 8. The centerline extraction results for three different images. (a) – (c) The directional lines are drawn manually on the original images. (d)-(f) The directional lines are drawn in gray based on the automatically extracted centerlines.

Weak Edge Leakage

Generally speaking region-based active contours will do better on weak edges but may have trouble in segmenting complex shapes. On the contrary it is shown above that edge-based active contour models can segment complex objects well when they are properly designed, but they need to overcome the drawback of weak-edge-leakage. In curve evolution, active contours can easily move across weak edge especially when there is a strong neighboring edge. Figure 9 illustrates the weak-edge-leakage problem, in which a weak edge is shown at the top-left corner of the image. The GVF segmentation result is obtained as shown in Figure 9(b), where the outmost black circle is the initial contour and the white contour is the final position of the curve. It is seen that the curve leak through the weak edge and conforms to the inner black circle.

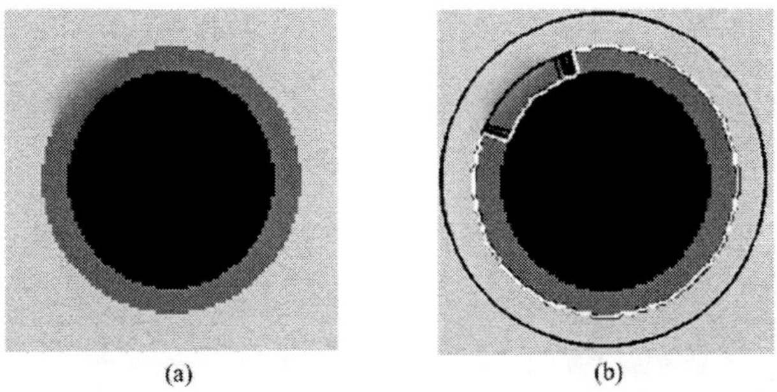

Figure 9. (a) A synthetic ring image with weak edge. (b) The convergence of the GVF active contour.

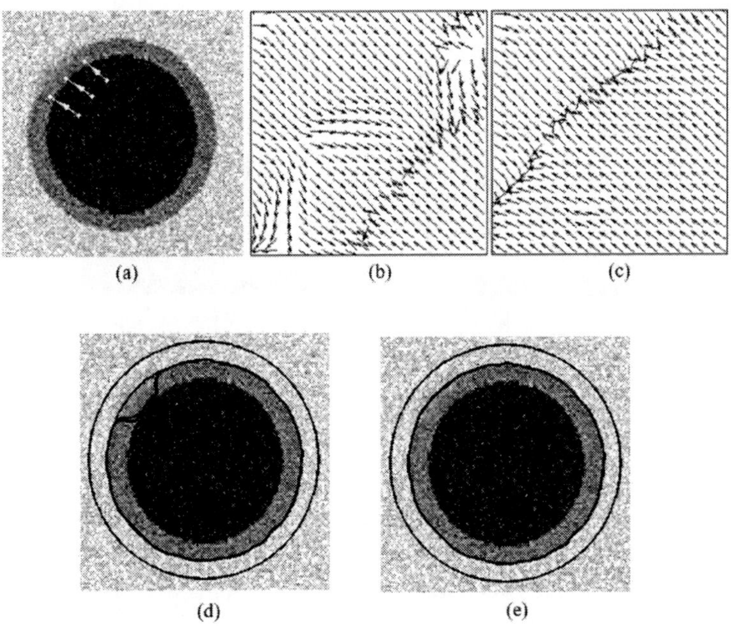

Figure 10. Segmentation of a synthetic ring image with weak edge. (a) Original synthetic image, on which three directional lines are used for the G&DVF. (b) and (c) The GVF field and the G&DVF field at the weak edge, respectively. (d) and (e) Segmentation results of the GVF active contour and the G&DVF active contour, respectively.

Below two methods for dealing with weak-edge-leakage are discussed. The first method is to utilize the G&DVF [15] described above. To prevent an active contour from moving across a weak edge, a straightforward solution is to push the contour back by adding an external force opposite to the curve evolution direction. This can be simply implemented by drawing directional lines perpendicular to the weak edge. Figure 10 shows an example of how to use the G&DVF for overcoming the weak-edge-leakage problem. In the noisy ring image shown in Figure 10(a), we add three directional lines to point to the weak edge to change the force field around the weak edge. Figures 10(b) and 10(c) respectively show the field vectors around the weak edge before and after adding the three lines. Without the three lines, the GVF field vectors around the weak edge point to the strong neighbor of the weak edge, which push the front of the GVF active contour to move across the weak edge and then to conform to its strong neighbor as shown in Figure 10(d). With three additional lines, the G&DVF field vectors around the weak edge point perpendicular to the weak edge, which make the front of the G&DVF active contour stay at the weak edge as shown in Figure 10(e).

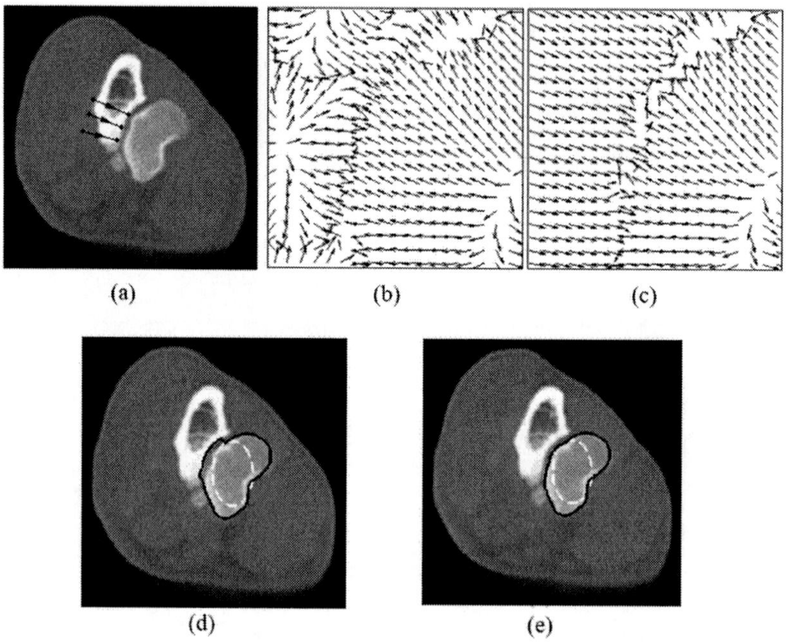

Figure 11. Segmentation of an elbow CT image with weak edge: (a) Original elbow image, on which three directional lines are used for the G&DVF. (b) and (c) The GVF field and the G&DVF field at the weak edge (pointed by the black directional lines), respectively. (d) and (e) Segmentation results of the GVF active contour and the G&DVF active contour, respectively.

Figure 11 shows another example on how to use the G&DVF for handling the weak-edge-leakage problem. Again the GVF field shown in Figure 11(b) moves the front of the GVF active contour across the weak edge, and makes the curve converge to the strong neighbor of the weak edge as shown in Figure 11(d). After adding three directional lines pointing to the weak edge for the G&DVF field as shown in Figure 11(a), the field vectors around the weak edge immediately are changed to point to the weak edge as shown in Figure 11(c). This makes the G&DVF active contour exactly converge to the elbow boundary as shown in Figure 11(e).

Anisotropic Virtual Electric Field

The anisotropic virtual electric (AVEF) field [21] is another method that can be used to overcome the weak-edge-leakage drawback of edge-based active contours. It is also based on changing the field forces to point perpendicularly to the weak edge when the field force is near the edge.

The AVEF is to extract the geometric information of the edge map and incorporate it into the original VEF model to change the field forces around the weak edge. The original VEF model was proposed by Park and Chung [3] to improve the performance of active contours by replacing the external force $-\nabla E_{ext}$ in the traditional active contour with the VEF, which treats each pixel in the edge map as an electric charge. Let $R_i = (x, y)$, $(i = 1, \cdots, N)$ denote the image domain and $f(x, y)$ denote the edge map, the VEF is computed as the gradient of the virtual electric potential (VEP) of the image edge map at location r as

$$F_{VEF}(r) = -\nabla P_{VEF}(r) = \sum_{k:R_k \neq r}^{N} \frac{e_k}{4\pi\varepsilon_0} \frac{r - R_k}{|r - R_k|^3}, \tag{15}$$

where the VEP is defined as

$$P_{VEF}(r) = \sum_{k:R_k \neq r}^{N} \frac{e_k}{4\pi\varepsilon_0} \frac{1}{|r - R_k|}. \tag{16}$$

In Eq. (16), e_k is called electric charge, defined as the intensity of the edge map, i.e., $e_k = f(R_k)$, and ε_0 is a positive constant that can be ignored.

The AVEF extracts the geometric information (the tangent direction of the edge map) as follows. Because the gradient of a smoothed gray-value image is approximately orthogonal to the tangent of the image edge, a geometric vector $\mathbf{v} = (v, u)$ is calculated to represent the geometric information for each virtual electric charge at position (x, y) given by

$$\begin{aligned} v(x,y) &= -\left[G_\sigma(x,y) * I(x,y)\right]_y \\ u(x,y) &= \left[G_\sigma(x,y) * I(x,y)\right]_x, \end{aligned} \tag{17}$$

where $[G_\sigma(x,y) * I(x,y)]_x$ and $[G_\sigma(x,y) * I(x,y)]_y$ are the spatial gradients of the smoothed image I along the x-axis and y-axis, respectively. Once the geometric vectors for all electric charges have been obtained, they are integrated into the VEP model as the anisotropic virtual electric potential (AVEP)

$$P_{AVEF}(r) = \sum_{k:R_k \neq r}^{N} \frac{e_k}{4\pi\varepsilon_0} \frac{\exp\left(\lambda |\langle r_k, \mathbf{v}\rangle|\right)}{|r - R_k|} \tag{18}$$

where r_k denotes $(r-R_k)/|r-R_k|$, which is the direction from electric charge e_k to location r, and $\langle\cdot,\cdot\rangle$ denotes inner product. The parameter λ is a nonnegative parameter, which controls the influence of the geometric vectors on the AVEP, and further on its corresponding force field. The intensity distribution of the AVEP generated by a single electric charge with geometric vector is anisotropic. The intensity of the potential, as distance increases from the electric charge, decreases more slowly along the direction of the geometric vector than along the vertical direction of the geometric vector. Therefore, the intensity of the AVEP is relatively enhanced along the image edge, even along weak image edges with strong neighbors. The AVEF is obtained by finding the gradient of the AVEP in Eq. (18) as:

$$F_{AVEF}(r) = -\nabla P_{AVEF}(r) = \sum_{k:R_k \neq r}^{N} \frac{e_k}{4\pi\varepsilon_0} \frac{\exp\left(\lambda\left|\langle r_k, \mathbf{v}\rangle\right|\right)\left[r_k - \lambda\frac{\langle r_k, \mathbf{v}\rangle}{|\langle r_k, \mathbf{v}\rangle|}\left(\mathbf{v} - \langle r_k, \mathbf{v}\rangle r_k\right)\right]}{|r-R_k|^2} \tag{19}$$

Because of the anisotropic property of the AVEF, the field forces will tend to point perpendicularly to image edges when they are near the edges, even near weak edges with strong neighbors. So they have the advantage of preventing active contours from moving across these weak edges. The ability of preventing boundary leakage of the AVEF is mainly determined by the parameter λ.

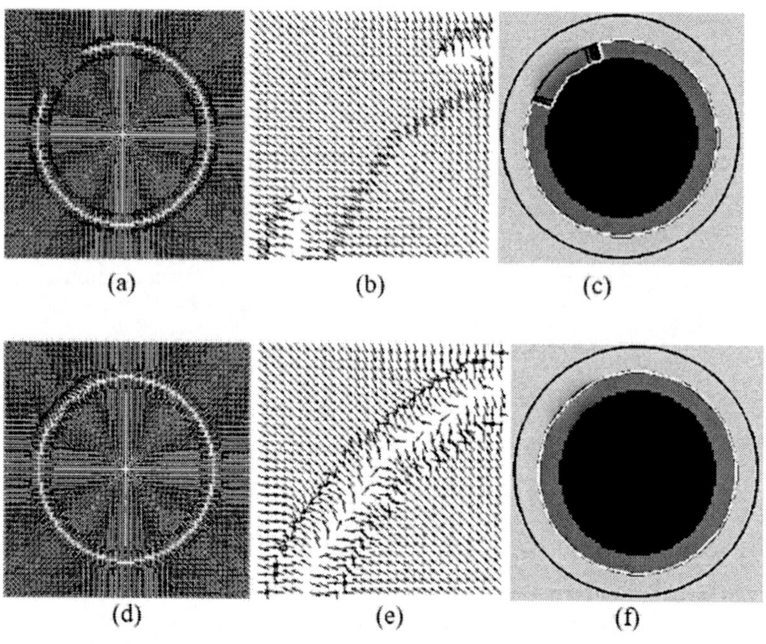

Figure 12. External force field, close-ups of the force fields at the weak edge and convergence of the active contours. (a)-(c) GVF active contour. (d)-(f) AVEF active contour. Note that the white lines represent the active contours at convergence.

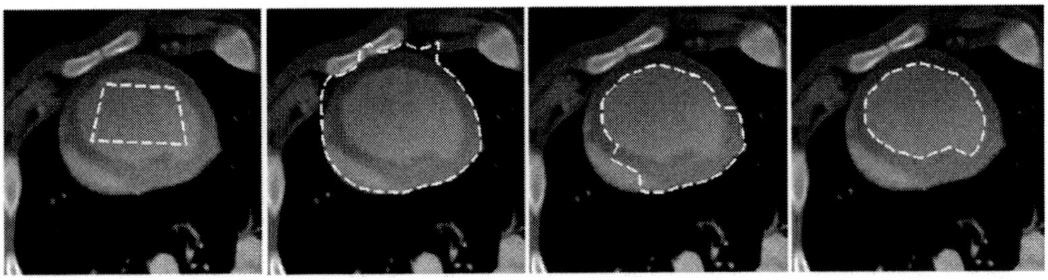

Figure 13. Segmentation of a cardiac CT image. (a) Original cardiac CT image with initialization. (b)-(d) Segmentation results of the GVF, VEF and AVEF active contour, respectively.

Figure 12 shows the effectiveness of the AVEF active contour in processing weak edge and the comparison with the GVF active contour. It is seen from Figure 12(b) that the field forces of the GVF at the weak edges point toward the strong neighboring edges, and therefore the GVF active contour leak through the weak edge and converge at the strong neighboring edge. On the contrary, the AVEF field forces point oppositely along the weak edges. So the AVEF active contour is prevented from moving across these edges and accurately conforms to the outside boundary of the synthetic ring image. Figure 13 show the segmentation results of a cardiac CT image with the GVF, VEF and AVEF, respectively, in which only the AVEF model can prevent the contour from moving across the weak edge.

Conclusion

Active contour models can often only produce satisfactory results for simple objects and have poor performance in segmenting images with complex object shapes, and they can leak through weak edge and converge to wrong object boundary. This chapter has reviewed several recently developed methods for overcoming these two problems. It has been shown that, when segmenting complex shapes, an effective approach is to properly guide the active contour's evolution direction either by detecting saddle and stationary points or providing directional prior manually or automatically. Geometric information of edge map and directional prior have been also shown to be capable of preventing active contours from moving across weak edges. A number of experimental results on image segmenattion have demonstrated the effectiveness of these active contour models.

References

[1] M. Kass, A. Witkin, and D. Terzopoulos, "Snakes: active contour models," *Int. J. Comput. Vis*. 1(4), pp. 321–331 (1988).

[2] L. D. Cohen and I. Cohen, "Finite-element methods for active contour models and balloons for 2-D and 3-D images," *IEEE Trans. Pattern Anal. Mach. Intell.* **15**(11), pp. 1131-1147 (1993).

[3] H. K. Park and M. J. Chung, "External force of snake: virtual electric field," *Electron. Lett.* **38**(24), pp. 1500-1502 (2002).

[4] X. Xie and M. Mirmehdi, 'Magnetostatic field for the active contour model: a study in convergence,' In Proceedings of the 17th British Machine Vision Conference, pp. 127-136, BMVA press, September (2006).

[5] K. W. Sum and Paul Y. S. Cheung, "Boundary vector field for parametric active contours," *Patt. Recog.* 40(6), pp. 1635–1645 (2007).

[6] B. Li and S. T. Acton, "Active contour external force using vector field convolution for image segmentation," *IEEE Trans Image Process.* 16(8), pp2096-106, 2007.

[7] C. Xu and J. L. Prince, "Generalized gradient vector flow external forces for active contours," *Signal Process.* 71(2), pp. 131-139 (1998).

[8] C. Xu and J. L. Prince, "Snakes, shapes, and gradient vector flow," *IEEE Trans. Image Process.* 7(3), pp. 359-369 (1998).

[9] J. Tang and S. T. Acton, "Vessel boundary tracking for intravital microscopy via multiscale gradient vector flow snakes," *IEEE Trans. Biomed. Eng.* 51(2), pp. 316- 324 (2004).

[10] C. H. Chuang and W. N. Lie, "A downstream algorithm based on extended gradient vector flow field for object segmentation," *IEEE Trans. Image Process.* 13(10), pp. 1379-1392 (2004).

[11] X. Xie and M. Mirmehdi, "RAGS: region-aided geometric snake," *IEEE Trans. Image Process.* 13(5), pp. 640-652 (2004).

[12] N. Ray and S. T. Acton, "Motion gradient vector flow: an external force for tracking rolling leukocytes with shape and size constrained active contours," *IEEE Trans. Med. Imaging* 23(12), pp. 1466-1478 (2004).

[13] H. Park, T. Schoepflin, and Y. Kim, "Active contour model with gradient directional information: directional snake," IEEE Trans. Circuits and Syst. *Video Technol.* 11 (2), pp. 252-256 (2001).

[14] J. Cheng, S.W. Foo, "Dynamic directional gradient vector flow for snakes," *IEEE Trans. Image Process.* 15(6), pp. 1563-1571 (2006).

[15] G. Zhu, S. Zhang, Q. Zeng, and C. Wang, "Gradient vector flow active contours with prior directional information," *Patt. Recog. Lett.,* 31(9), pp.845-856 (2010).

[16] J. Ning, C. Wu, S. Liu, and S. Yang, "NGVF: An improved external force field for active contour model," *Patt. Recog. Lett.* 28(1), pp. 58-63 (2007).

[17] X. Xie and M. Mirmehdi, "Geodesic colour active contour resistent to weak edges and noise," In Richard Harvey and Andrew Bangham, editors, Proceedings of the British Machine Conference, pages 41.1-41.10. BMVA Press, September 2003

[18] G. Zhu, S. Zhang, X. Chen, and C. Wang, "Novel Gradient vector flow-based balloon force for active contours," *J. of Electronic Imaging* 18(2), pp. 023007-1 - 8 (2009).

[19] N. Paragios, O. Mellina-Gottardo, and V. Ramesh, "Gradient vector flow fast geometric active contours," *IEEE Trans. Pattern Anal. Mach. Intell.* 26(3), pp. 402-407 (2004).

[20] S. Zhang and J. Zhou, "Centerline Extraction for Image Segmentation using Gradient and Direction Vector Flow Active Contours," *J. of Sig. and Info. Proc.* 4(4), pp. 407-413 (2013).

[21] G. Zhu, S. Zhang, Q. Zeng, and C. Wang, "Anisotropic virtual electric field for active contours," *Patt. Recog. Lett.,* 29 (11), pp. 1659-1666 (2008).

In: Digital Signal Processing (DSP)
Editor: Juan Zhang

ISBN: 978-1-63485-168-8

Chapter 3

AVERAGE FILTERING: THEORY, DESIGN AND IMPLEMENTATION

Davis Montenegro[*] and Javier Gonzalez[†]
Santo Tomas University, Bogota, Cundinamarca, Columbia

ABSTRACT

Digital filtering is a set of algorithms based on differential equations. The simplest algorithm within this set is the Finite Impulse Response (FIR) filter. This only requires the input samples to generate the filtered output, avoiding the feedback loops. FIR filters can be designed and implemented according to the need, and because of their simplicity have found application on many fields including Real-Time systems. A special implementation of a low pass algorithm is the averaging filter. It calculates the output sample using the average from a finite number of input samples. The averaging filter is used in situations where is necessary to smooth data that carrying high frequency distortion. The main aim of this chapter is the exposition of the theory, implementation and application of the average filtering. This chapter also presents the application of the average filteringin two study cases: electrophysiological signals and electrical power signals.

Keywords: average, digital signal processing, filtering, low order, real-time

INTRODUCTION

Nowadays data acquisition using digital interfaces is the preferred technique for describing and analyzing physical phenomenon. Depending on the primary element (sensor) and the environmental conditions a wide range of high frequency noise can be part of the acquired waveform.

[*] E-mail: davismontenegro@usantotomas.edu.co (Davis Montenegro).
[†] E-mail: javiergonzalezb@usantotomas.edu.co (Javier Gonzalez).

This situation demands a cleaning stage where the involved noise gets reduced after the acquisition process. With this process it is expected that further stages will receive clean data for performing analysis. As result, the accuracy of the obtained information for describing the phenomenon gets improved.

But not all applications look for the same goal. There are fields of application where it is required to eliminate signal components which are harmonics of the desired waveform. Once the undesired signal components are separated these can be used for performing control actions for eliminate them.

The cases mentioned above can be found in instruments, signal coditioning equipment, control equipment, among others. These applications have a common request: Fast algorithms for implementation with several hardware architectures for real-time operation.

These kind of problems can be addressed by using digital filtering. Digital filtering is a set of algorithms based on differential equations. The simplest algorithm within this set is the Finite Impulse Response (FIR) filter. This requires only the input samples to generate the filtered output, avoiding the feedback loops. FIR filters can be designed and implemented according to the need, and because of their simplicity have found application on many fields including Real-Time systems.

A special implementation of a low pass filters is the averaging filter. It calculates the output sample using the average value from a finite number of input samples. The averaging filter is used in situations where data carrying high frequency distortion need to be smoothed.

The aim of this chapter is the exposition of the theory, implementation and application of the average filtering. This chapter also presents the application of the average filtering in two study cases: electrophysiological signals and electrical power signals.

THE AVERAGING FILTER

Theoretical Background

A digital filter is a discrete system designed for processing data stored in arrays. Mathematically, a digital filter can be described using a differential equation as shown in (1).

$$\sum_{k=0}^{L-1} V_k y(n-k) = \sum_{k=0}^{L-1} W_k x\,(n-k) \tag{1}$$

An Alternative form is the Finite Impulse Response (FIR) filter. This uses only the actual and previous input for performing the filtering action. This form avoids the uses of previous outputs, thus reducing the computational burden when calculating the output y(n) as shown in (2).

$$y[n] = \sum_{k=0}^{L-1} W_k x(n-k) \tag{2}$$

A digital FIR structure with a particular function is the average filter. This Filter attenuates higher frequency components and smooths the signal. The average filter is shownin (3), where the variable L is de order of average filter.

$$y[n] = \frac{1}{L}\sum_{k=0}^{L-1} x(n-k) \quad (3)$$

These filters have the following characteristics [1]:

1. The transfer function only has one constant term. The other terms are always the coefficients of the shift registers where the inputs are stored. Additionally, all the poles are in the origin of the Z plane. This guarantees the stability of the filter because the impulse response counts with a finite number of terms.
2. The operations of the FIR filter involves only multiply the inputs by its coefficients for being accumulated. The implementation of this filter is very simple.
3. Given the format of the FIR filter its coefficients can be found to obtain the desired response. This filter can be configured as low-pass, band-pass and high pass.

There are several methods for obtaining the coefficients of this filter. These methods use the Fast Fourier Transform, time windows, frequency sampling among others [2]. Some of them and their practical application are shown as follows.

Design Algorithms

The frequency response of the average filter depends of the L value according to (2) and (3). Using the following code in MATLAB, it is possible estimate the frequency response.

```
l = [2 4 8 16 32 64];
N = length (l);
for i = 1:N
L = l (i);
B = (1/L)* ones (1, L);
[H, F] = freqz (B, 1, 200, Fs);
A (i,:) = abs (H);
end;
```

With this algorithm it is possible to calculate the response of the filter FIR in the frequency domain for different values of L. In Figure 1 the response for L = 4, 8 and 16 are shown.

In Figure 1 the operational characteristics of the filter can be appreciated such as the Gibbs oscillations, the ripple at the cut off frequency, among others. Then, for testing the performance of the filter with different values of L a waveform is built using a sample frequency of 500 Hz. This waveform is composed by 4 frequencies as shown in the following code:

```
Fs = 500;
Ts = 1/Fs;
n = 1:500;
t = (n-1)*Ts;
```

*s = cos (2*pi*5*n*Ts);*
*r = cos (2*pi*13*n*Ts) + cos (2*pi*72*n*Ts) + cos (2*pi*130*n*Ts);*
x = s + r;

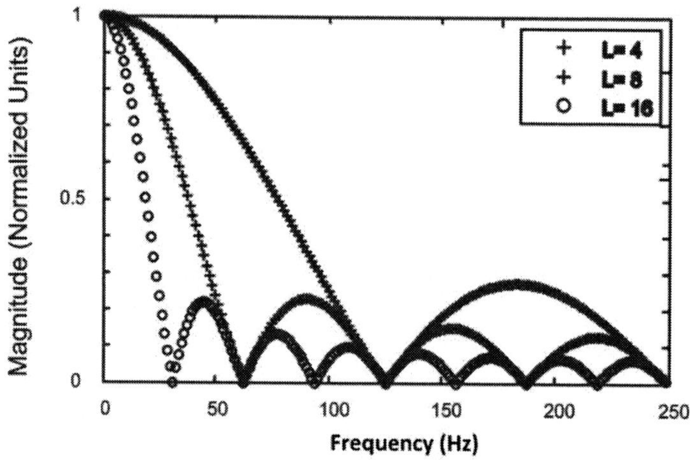

Figure 1. Response of the FIR filter for L = 4, 8 and 16.

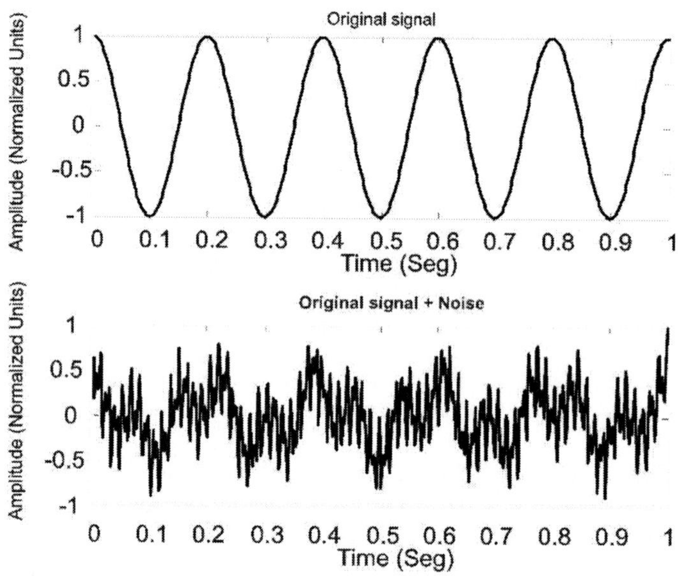

Figure 2. Clean waveform (Up) and Contaminated waveform (Bottom).

As result, the signal x (n) is composed by two parts: the clean and desired waveform and the noise. The clean waveform is s(n) and the noise is represented by r(n). Both components are shown in Figure 2.

The contaminated waveform x(n) will be entered to the filter and the output signal y(n) will be plotted. For this example 4, 8, 16 and 32 coefficients (L) are proposed for designing 4 different filters. This procedure is made using the following code:

```
l= [4 8 16 32];
N = length (l);
for I = 1:N
L = l(i);
b = (1/L)* ones (1, L);
y (i,:) = filter(b, 1, x);
end;
```

The obtained outputs for each filter are shown on Figure 3. In this Figure is notable the low-pass effect of the filter and the smooth reached depending on the order of the filter. Additionally, it is evident that when the order of the filter gets increase the performance of the filter is better. However, the increment in the filter order means a higher computational burden. It is then when a balance must be found between computational burden and the filter response. Another consideration for the FIR filter is the phase shifting when increasing the order of the filter.

For implementing this filter in different languages a simple coding is required. One technique for a simple implementation is using the concept of convolution [3]. According to its definition a convolution can be described as shown in (4).

$$y(n) = x(n) * h(n) \tag{4}$$

In this case h(n) is a discrete function and its coefficients corresponds to the value of L. This discrete function is described in (5) where it is an array of size L filled with ones divided by L.

$$h(n) = \left(\frac{1}{L}\right)[1_0 1_2 \dots \dots . 1_{L-1}] \tag{5}$$

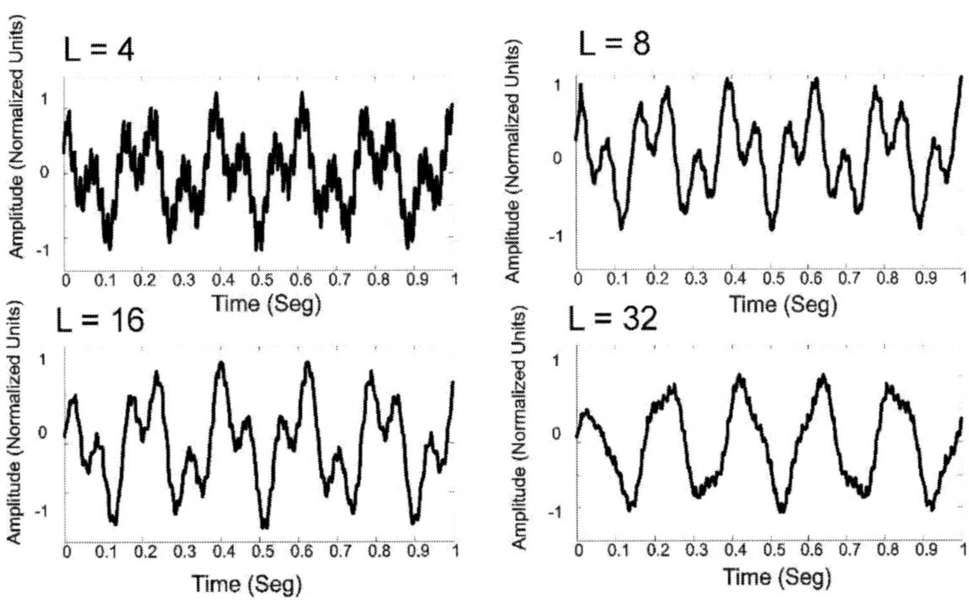

Figure 3. Signal y(n) obtained from different versions of the average filter.

This implementation can be performed using C++ in a simple way. The following code can be used for implementing a simple average filter in a regular computer. The input samples are stored in the shift array x and the result of the convolution loop is placed in the variable c.

```
for (n = L + 1; n <= N; n ++)
{c = 0;
for (k = 1; k <= L, k ++)
c = c + x(n-k);
y (n) = c/L;}
```

But sometimes the time becomes a major concern. In these cases users would like to implement this kind of filters using hardware for covering a wider spectrum. The preferred technologies for this are FPGAs. Figure 4 shows the implementation of an average filter based on shift registers which can be described in hardware using VHDL.

In Figure 4, the shift registers Rn are used for storing the samples taken with the analog to digital converter (ADC) in time. The register B0 is used for storing the result of operating these samples according to (5) using bit right shifting. As result, the filtered signal will be found after operate all the coefficients of the filter corresponding to the shift register length.

This structure can be implemented using VHDL with the following code:

```
identity FILTER is
Port (X: in STD_LOGIC_VECTOR (7 downto 0);
C: in STD_LOGIC;
Y: out STD_LOGIC_VECTOR (7 downto 0);
end tarjeta;

architecture Behavioral of FILTER is
signal R0:std_logic_vector (7 downto 0);
signal R1:std_logic_vector (7 downto 0);
signal R2:std_logic_vector (7 downto 0);
signal R3:std_logic_vector (7 downto 0);
B0:std_logic_vector (7 downto 0);
B1:std_logic_vector (7 downto 0);

begin
process (C)
begin
if C = '0' and C'event then
X < = not A;
R0 <= not (A);
R1 <= R0;
R2 <= R1;
R3 <= R2;
end if;
```

```
end process;
B0 <= R0 + R1 + R2 + R3;
B1 (7) <= '0';
B1 (6) <='0';
B1 (5) <= B0 (7);
B1 (4) <= B0 (6);
B1 (3) <= B0 (5);
B1 (2) <= B0 (4);
B1 (1) <= B0 (3);
B1 (0) <= B0 (2);
Y <= B1;
end Behavioral;
```

This code can be used for reproducing the proposed filter within a FPGA. This device can then be used for several purposes.

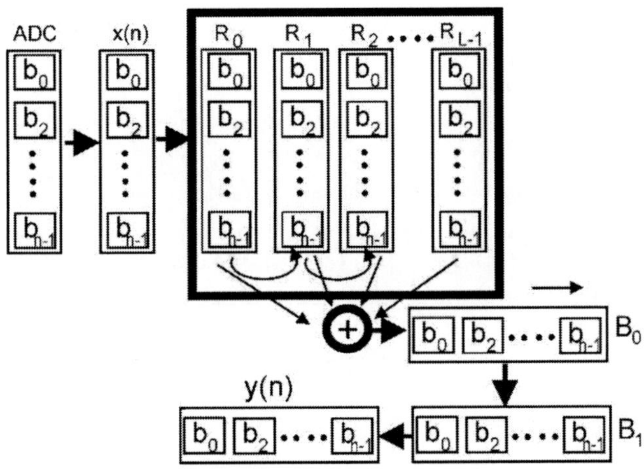

Figure 4. Average filter based on shift registers.

Applications in the Time and Frequency Domains

Many applications of the average filter can be found in the time and frequency domains. Sometimes, the input signals have baseline components for which modern technology is not immune. The problem with this kind of components is that they contaminate the low-frequency components of the studied signal [4]. In the Followings lines of code for MATLAB, it is possible simulate a discrete signal with a base line component.

```
fs = 500;
Ts = 0.001;
n = 1:1000;
t = (n-1)*Ts;
s1 = sin (2*pi*10*n*(1/fs));
s2 = sin (2*pi*15*n*(1/fs));
```

*s3 = sin (2*pi*20*n*(1/fs));*
*s4 = sin (2*pi*25*n*(1/fs));*
*s5 = sin (2*pi*30*n*(1/fs));*
x = s1 + s2 + s3 + s4 + s5;
*v = 2*sin (2*pi*0.2*n*(1/fs));*
s = x + v;

This code will generate a waveform based on sinusoidal signals. The baseline of a waveform can be described as a background that surrounds the important data of the composed signal [5]. This background represents a source of noise that can affect the accuracy of the information contained in the time window.

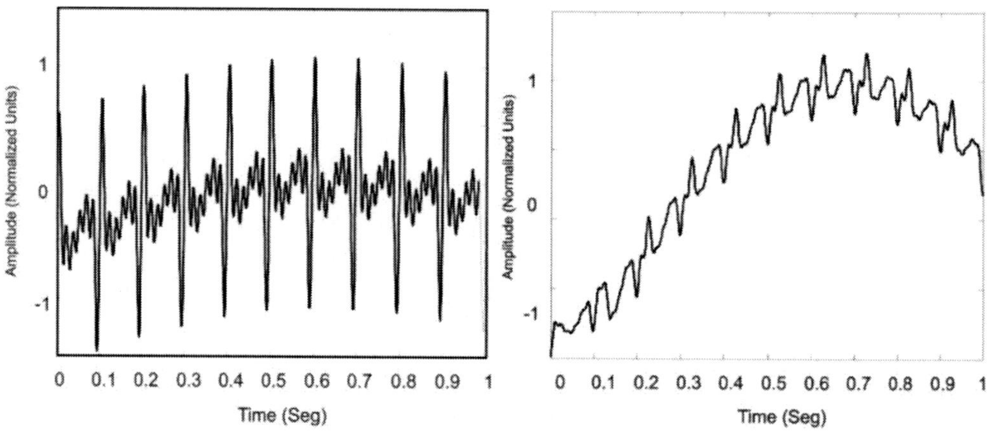

Figure 5. Waveform generated containing baseline components (left) and the baseline (right).

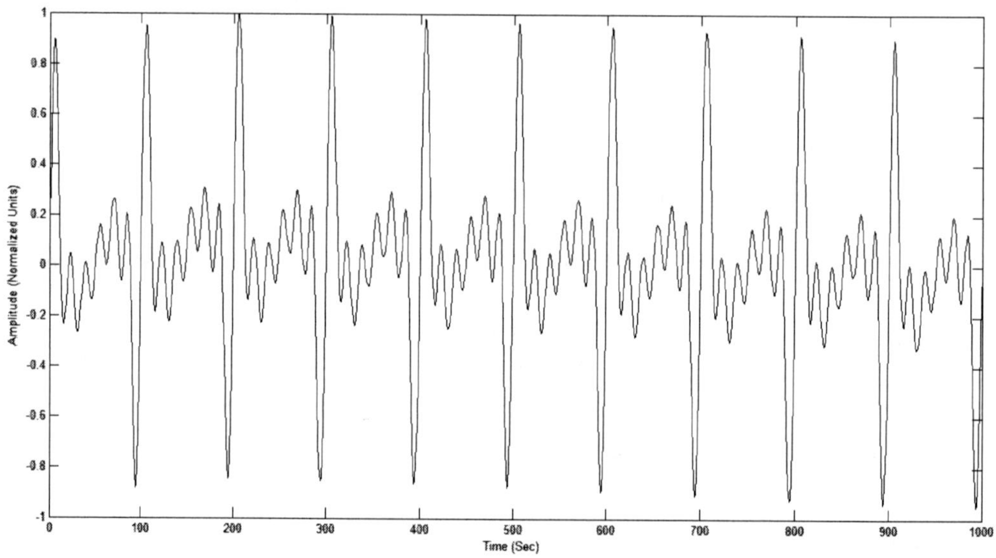

Figure 6. Waveform obtained after the baseline removal.

The waveform generated with the code presented above is shown in Figure 5. In this Figure it is possible to appreciate the concepts of baseline mentioned previously.

For isolating the baseline from the composed waveform an average filter can be implemented using 128 memory blocks (L). The code for implement this filter and extract the baseline as show in Figure 5 (right) is as follows:

```
L = 128;
b = (1/L)*ones (1, L);
y = filter (b, 1, s);
```

This simple implementation allows by generating a group of coefficients with the same value to characterize the useless data. In this particular case this modeled noise can be used for removing it from the original signal before perform extract the desired information. In Figure 5 if the signal at the right is algebraically subtracted from the signal at the left the base line components gets eliminated. The resulting waveform is shown in Figure 6.

These procedures are common in several applications. Two examples of these applications are going to be presented as follows.

Study Case: Electrophysiological Signals

The electrocardiogram is an electrophysiological signal that describes the electrical behavior of the human heart. The electrocardiographic signal (ECG) requires several measurements from different parts of the patient's body (biomedical instrumentation). However, the presence of electrical equipment adds disturbances that affect the instrumentation system.

The ECG's amplitude is around 1mV and its bandwidth is between 0.5 and 100 Hz. Figure 9 shows an electrocardiographic signal taken from the signals data base Physionet [6]. The proposed signal and its components in the frequency domain are shown in Figure 7.

This signal is going to be altered by adding noise corresponding to the noise generated by the proximity of power lines, which is common in ECG signals once they are acquired. The MATLAB code used for generate the new waveform is presented as follows:

```
load signal
fs = 500;
Ts = 1/fs;
N = 1000;
n = 1:N;
t= (n-1)/fs;
r = 0.5*sin (2*pi*60*n*Ts);
x1 = s + r;
```

The obtained waveform is shown in Figure 8. This new version of the original signal is closer to a real scenario where different factors, such as power lines, can add undesired data to the time window.

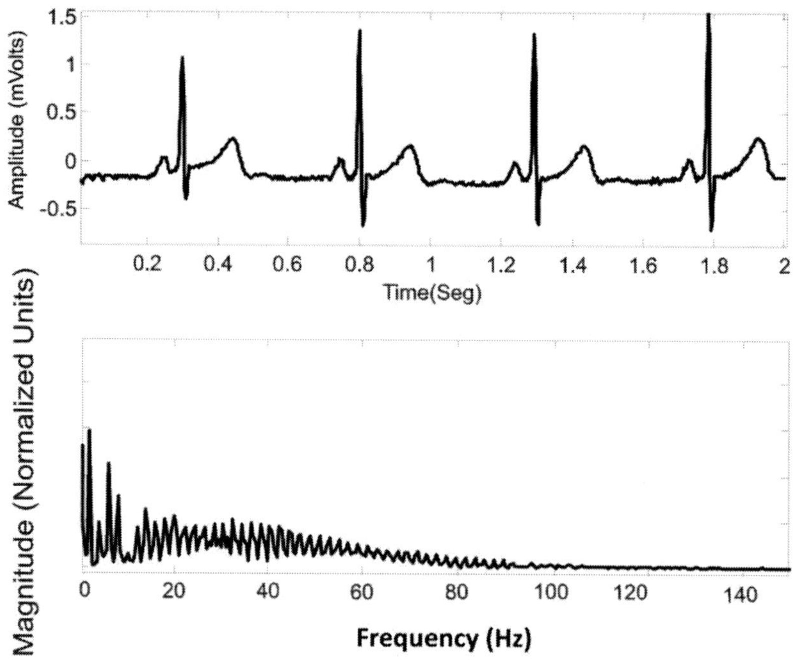

Figure 7. Proposed ECG signal and its spectrum.

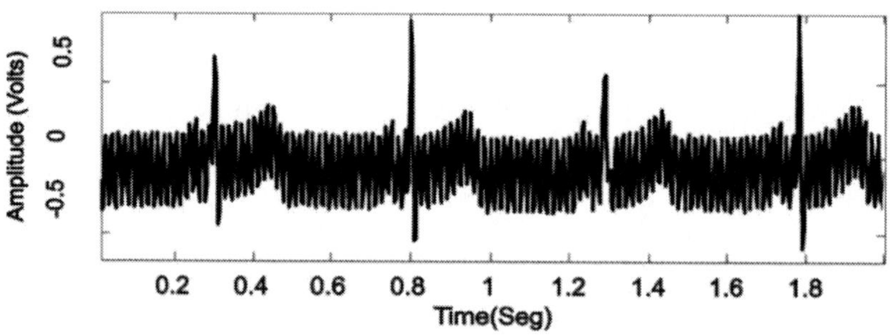

Figure 8. Distorted ECG signal by the proximity of power lines.

Then following the procedure presented above the average filter can be obtained by using a simple code in MATLAB. In this case a low order is selected for implement the filter (L = 8); the aim is to verify how with low orders the filter reach a good performance for this kind of tasks. The implemented code is as follows:

```
L = 8;
b = (1/L)*ones (1, L);
y = filter (b, 1, x1);
y = y/max (y);
```

As result the noise added initially becomes reduced significantly as shown in Figure 9. This demonstrates that the average filter results highly efficient for working as low-pass filter and eliminate the undesired components of the studied signal. Additionally the computational burden is not demanding for Real-time application.

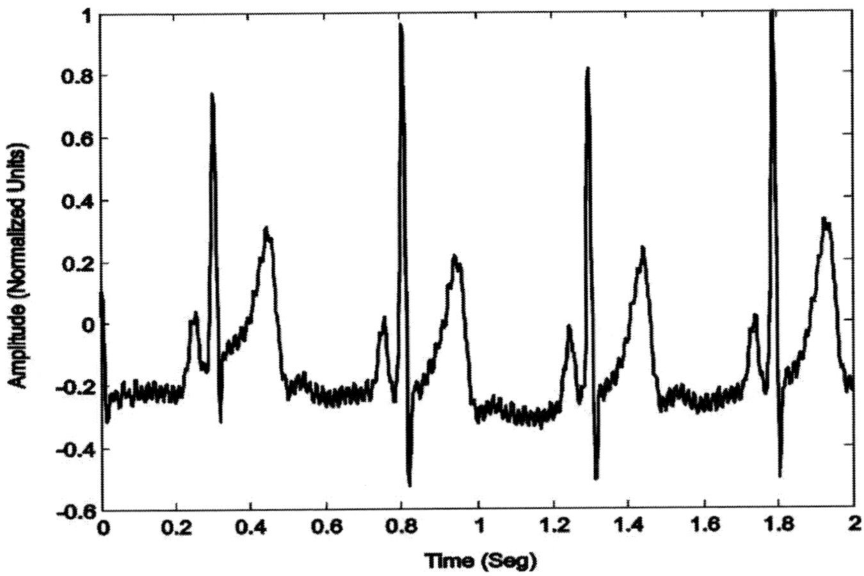

Figure 9. Filtered ECG signal using average filtering with L = 8.

Study Case: Active Filters for Power Conditioning

In power systems the connection of non-conventional loads such as power inverters, electronic power sources, Electric Vehicles, among others, has led a big workforce interested in power quality issues [7, 8]. The power quality disturbances are already identified and documented in international standards [9, 10]; these standards define the characteristics of the disturbances for monitoring and control.

Harmonic distortion is a very common disturbance and there is a big interest in reduce it in power systems [11, 12] for building the smart grid. Using active power filters the harmonics can be modeled and eliminated from the power system by injecting or draining energy. Figure 10 shows the general schematic of an Active filter for power conditioning connected in shunt.

The control of the semiconductors for injecting the correcting signal is performed by the active filter controller. This structure was proposed by Akagi et al. [13] for applying the instantaneous power theory and eliminating the harmonics present in the power lines [14].

The interesting about this application is that works with only one sample for infer the operation of the semiconductor bridge. As result, energy can be dragged and stored in the capacitors or injected when it is necessary.

In this special application the use of average filtering results adequate. First the voltage and current signal samples are acquired and transformed from the *a, b, c* plane to the *α, β, 0* plane using Clark's transformation as shown in (6).

$$\begin{bmatrix} I_{ca} \\ I_{cb} \\ I_{cc} \end{bmatrix} = \sqrt{\frac{2}{3}} \begin{bmatrix} 1/\sqrt{2} & 1 & 0 \\ 1/\sqrt{2} & -1/2 & \sqrt{3}/2 \\ 1/\sqrt{2} & -1/2 & -\sqrt{3}/2 \end{bmatrix} \begin{bmatrix} -i_0 \\ i_{c\alpha} \\ i_{c\beta} \end{bmatrix} \tag{6}$$

Then, terms associated with 0 sequence (ground) can be discarded because they do not contribute to the non-active power [15]. With this reduces vector the reference frame can change from *α, β, 0* to the *p, q* plane. This new frame separates the active power from the non-active power, which represents the undesired components of the power signal (harmonics). This transformation is performed as follows:

$$\begin{bmatrix} i_{c\alpha} \\ i_{c\beta} \end{bmatrix} = \frac{1}{V_\alpha^2 + V_\beta^2} \begin{bmatrix} V_\alpha & -V_\beta \\ V_\beta & V_\alpha \end{bmatrix} \begin{bmatrix} -\tilde{p} + \Delta\bar{p} \\ -\bar{q} - \tilde{q} \end{bmatrix} \quad (7)$$

In (7) V_α and V_β were obtained using the same procedure presented in (6). These equations are invertible making possible to go from different reference frames for control purposes. After equation (7) is operated the composed power signal $-\tilde{p} + \Delta\bar{p}$ can be filtered using the current sample.

This consideration means that the action of the filter is performed after a number of iteration equal to the length of the filter's shift register. So if the filter has a length of 8 registers means that only after 8 iterations the filter will be initialized. This initialization will guarantee that all the coefficients of the filter are different to 0 and the filtering action will be performed.

After the filtering the new term *p* is transformed in inverse for determining the current value for injecting/drain from the power system. The injection/drain sequence of the semiconductor bridge is made by applying a hysteresis algorithm that delivers the triggering sequence for each phase.

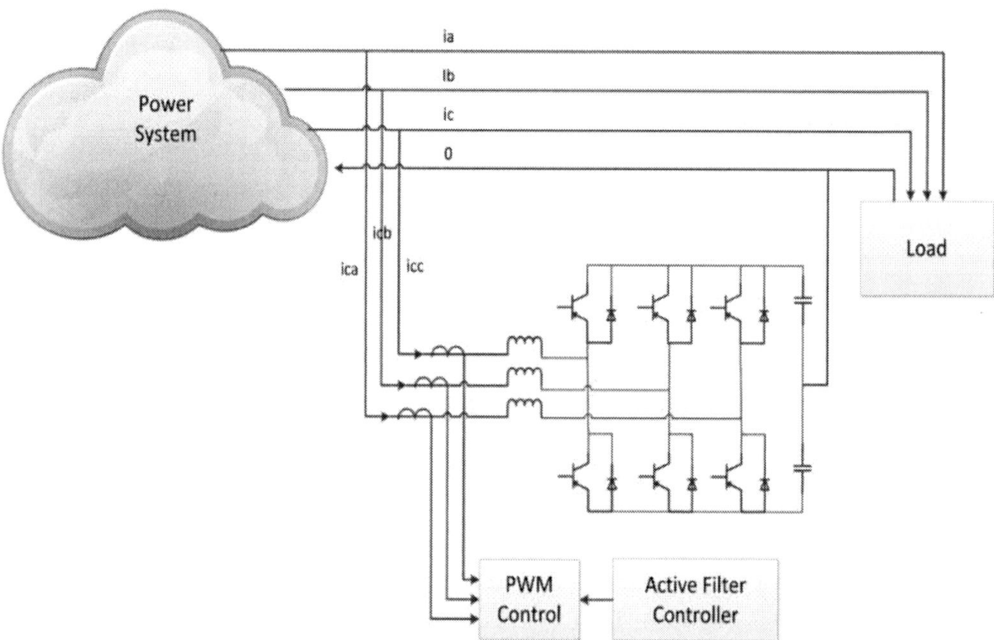

Figure 10. General schematic of active filters for power conditioning.

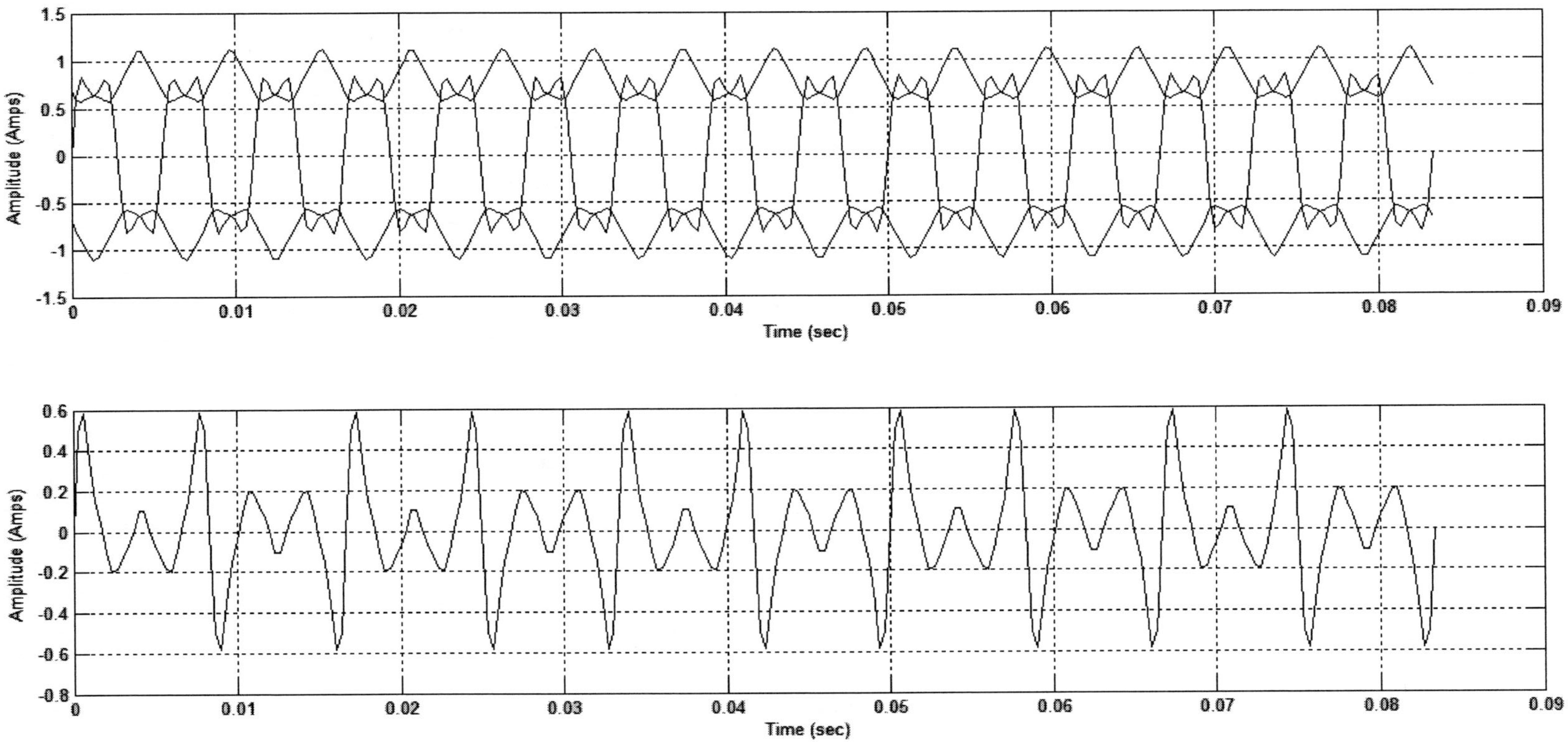

Figure 11. Distorted current waveforms (top) and filtered non active current (bottom).

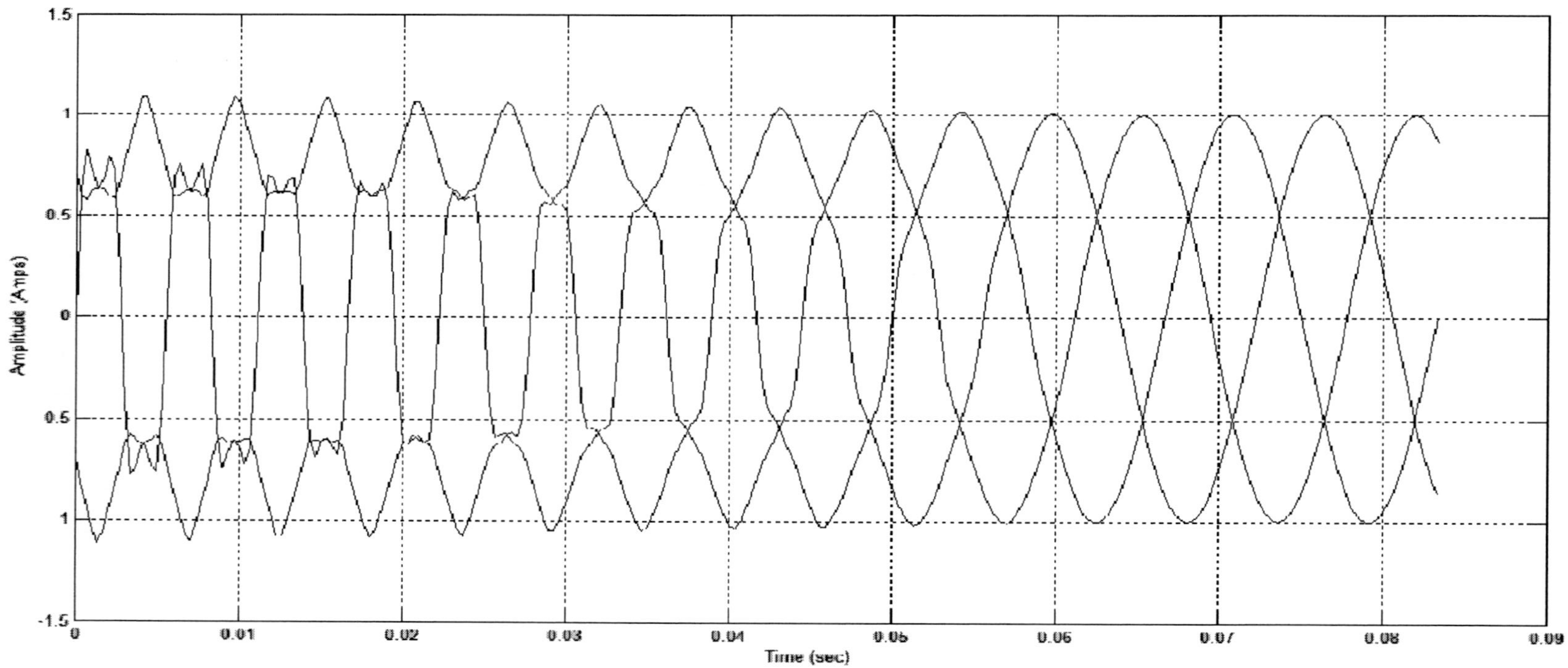

Figure 12. Obtained current signal (phase a) when the filter is operating.

The following MATLAB code generates the current waveforms of a three phase 6 pulse inverter; this device can be found in some Electric Vehicles. The X/R ratio of the system is quite low making the total harmonic distortion (THD) of the voltage signal lower than 1.

```
%Manitudes of odd harmonics and fundamental (Inverter IEEE 519)
Harm = [1 0.1378 0.30 0.12 0.089 0.056 0.044];
Fs = 3000; %Sample Frequency S/s
FFund = 60; %Fundamental Freq (60 Hz)
Per = 5; %Periods
Long = (Fs/FFund)*Per + 1; %length of the vectors
N = 0:Long-1; %Samples equivalent to "Per" periods
theta = 2*pi/3; %Phase angle
Ic = zeros (3, Long);
Tam = size (Harm);
% Generate the 3 phase current waveforms
for Fr = 1:Tam(2), %Tam(2),
Ic = Ic + [Harm (Fr)*sin((Fr*2-1)*2*pi*FFund*n/Fs);
Harm (Fr)*sin ((Fr*2-1)*(2*pi*FFund*n/Fs + theta));
Harm (Fr)*sin ((Fr*2-1)*(2*pi*FFund*n/Fs-theta))];
end;
t = n/Fs;
plot (t, Ic);
grid on;
xlabel ('Time (sec)');
ylabel('Amplitude (Amps)');
```

The generated waveforms are shown in Figure 11 (up). The magnitude of these signals is normalized and represents a demand big enough for impact negatively the power system. After performing the plane transformation of the current and voltage waveforms the average filter is applied. Then the non-active power is transformed into the harmonic current shown in Figure 11 (bottom). The resulting current waveform when the filter operates is shown in Figure 12.

CONCLUSION

This chapter has presented an introduction to the theory, design and implementation of average filters. The characteristics of these filters make of them easy for designing and implement. They are widely applied in several fields due their suitability for Real-Time execution. Two of these applications have been presented and also some codes for simulating them.

REFERENCES

[1] Tan, L. (2007). *Digital Signal Processing: Fundamentals and Applications*: Elsevier Science.

[2] Proakis, J. G. and Ingle, V. (1997). *Digital Signal Processing Using MATLAB V4.0*. Boston, United States of America: PWS Publishing Company.

[3] Proakis, J. G. and Manolakis, D. G. (1995). *Digital Signal Processing*. New Jersey, U.S.A.: Prentice- Hall, Inc.

[4] DeLuca, C. J., Gilmore, L. D., Kuznetsov, M. and Roy, S. H. (2010). Filtering the surface EMG signal: Movement artifact and baseline noise contamination. *Journal of Biomechanics, 43*, 1573-1579. doi: 10.1016/j. jbiomech.2010.01.027.

[5] Chouhan, V. S. and Mehta, S. S. (2007, 5-7 March 2007). *Total Removal of Baseline Drift from ECG Signal*. Paper presented at the International Conference on Computing: Theory and Applications, 2007. ICCTA '07.

[6] NIBIB. (2015). *Physionet: The research resource for complex physiological signals*. Retrieved January, 2015, from http://www. physionet.org/.

[7] Arritt, R. F. and Dugan, R. C. (2011). Distribution System Analysis and the Future Smart Grid. *IEEE Transactions on Industry Applications, 47*(6), 2343-2350. doi: 10.1109/TIA.2011.2168932.

[8] Zavoda, F. (2010). *Advanced Distribution Automation (ADA) Applications and Power Quality in Smart Grids*. Paper presented at the China International Conference on Electricity Distribution, Shanghai.

[9] IEEE Recommended Practice for Monitoring Electric Power Quality. (2009) *IEEE Std 1159-2009 (Revision of IEEE Std 1159-1995)* (pp. c1-81).

[10] IEEE Recommended Practices and Requirements for Harmonic Control in Electrical Power Systems. (1993). *IEEE Std 519-1992*, 0_1. doi: 10.1109/IEEESTD.1993.114370.

[11] Aredes, M. and Watanabe, E. H. (1995). New control algorithms for series and shunt three-phase four-wire active power filters. *IEEE Transactions on Power Delivery, 10*(3), 1649-1656. doi:10.1109/61. 400952.

[12] Bollen, M. H. J., Gu, I. Y. H., Santoso, S., McGranaghan, M. F., Crossley, P. A., Ribeiro, M. V. and Ribeiro, P. F. (2009). Bridging the gap between signal and power. *Signal Processing Magazine, IEEE, 26*(4), 12-31. doi: 10.1109/MSP.2009.932706.

[13] Akagi, H., Kanazawa, Y. and Nabae, A. (1984). Instantaneous Reactive Power Compensators Comprising Switching Devices without Energy Storage Components. *IEEE Transactions on Industry Applications, IA-20*(3), 625-630. doi: 10.1109/TIA. 1984.4504460.

[14] Akagi, H., Watanabe, E. H. and Aredes, M. (2007). *Instantaneous Power Theory and Applications to Power Conditioning*: Wiley.

[15] Watanabe, E. H., Stephan, R. M. and Aredes, M. (1993). New concepts of instantaneous active and reactive powers in electrical systems with generic loads. *IEEE Transactions on Power Delivery, 8*(2), 697-703. doi: 10.1109/61.216877.

In: Digital Signal Processing (DSP)
Editor: Juan Zhang

ISBN: 978-1-63485-168-8

Chapter 4

THE REGULARIZED BANDPASS FILTER

Weidong Chen*

Department of Mathematics, Park University,
Parkville, Missouri, US

ABSTRACT

In this chapter, the bandpass filter is discussed in the noisy case. And a regularized bandpass filter is presented. The convergence property of the regularized bandpass filtering algorithm is proved in theory and tested by numerical results.

Keywords: noise, bandpass filter, regularization

INTRODUCTION

Filtering is widely applied in engineering [1-5]. In this paper, the problem of the bandpass filtering is analyzed in theory and by examples in detail. A regularized bandpass filtering algorithm is presented with the proof of the convergence property and numerical results. First, we describe the band-limited signals $f \in L^1(\mathbf{R})$. The details can be seen in [7].

Definition: A function $f \in L^1(\mathbf{R})$ is said to be Ω-band-limited if

$$\hat{f}(\omega) = 0, \ \forall \omega \notin [-\Omega, \Omega].$$

Here $\hat{f}$ is the Fourier transform of f:

$$F(f)(\omega) = \hat{f}(\omega) = \int_{-\infty}^{+\infty} f(t) e^{i\omega t} dt, \ \omega \in \mathbf{R} \tag{1}$$

* E-mail: chenw@ksu.edu.

We then have the inversion formula:

$$F^{-1}(\hat{f})(t) = f(t) = \frac{1}{2\pi}\int_{-\Omega}^{\Omega} \hat{f}(\omega)e^{-i\omega t}d\omega, \text{ a.e. } t \in \mathbf{R}.$$

In many practical problems, the signal $f(t)$ is noisy:

$$f(t) = f_E(t) + \eta(t) \tag{2}$$

where $\eta(t)$ is the noise

$$|\eta(t)| \le \delta \tag{3}$$

and $f_E \in L^1$ is the exact band-limited signal.

In this paper, we will consider the problem bandpass filtering:

$$f_L(t) = \int_{-\infty}^{\infty} f(u)\frac{\sin\Omega(t-u)}{\pi(t-u)}du, \text{ a.e. } t \in \mathbf{R}. \tag{4}$$

If the signal is noisy however, the filter is not reliable. We will give an example to show that the noise can become very large after the bandpass filtering process.

So this filter is not reliable in the noisy case. And a regularized bandpass filtering algorithm will be presented.

In section 2 we give the property of the bandpass filter. A regularized filtering algorithm and the proof of its convergence are in section 3.

The numerical results of some examples are given in section 4. Finally, the conclusion is given in section 5.

THE PROPERTY OF THE BANDPASS FILTER

In this section, we discuss the property of the bandpass filter.

Example: Assume the noise is $\eta(t) = \varepsilon\, sgn\{\sin\Omega(t_0 - t)/\Omega(t_0 - t)\}$ where t_0 is a given point in the time domain and ε is close to zero. Then the noise signal after the filtering is

$$\eta_L(t) = \int_{-\infty}^{\infty} \eta(u)\frac{\sin\Omega(t-u)}{\Omega(t-u)}du = \varepsilon\int_{-\infty}^{\infty}\frac{\sin\Omega(t-u)}{\Omega(t-u)}sgn\left[\frac{\sin\Omega(t_0-u)}{\Omega(t_0-u)}\right]du.$$

We can see that $|\eta(t)| \le \varepsilon$. However, the noise at $t = t_0$ after the filtering is

$$\eta_L(t_0) = \varepsilon \int_{-\infty}^{\infty} \left| \frac{\sin \Omega(t_0 - u)}{\Omega(t_0 - u)} \right| du = \infty.$$

Also at any point $t = t_0 + k\pi / \Omega$, $k \in \mathbf{Z}$

$$\eta_L(t_0 + k\pi / \Omega) = \varepsilon \int_{n=-\infty}^{\infty} (-1)^k \frac{\sin \Omega(t_0 - u)}{\Omega(t_0 + k\pi / \Omega - u)} sgn \left[\frac{\sin \Omega(t_0 - u)}{\Omega(t_0 - u)} \right] du$$

$$= (-1)^k \varepsilon \int_{-\infty}^{\infty} \frac{\Omega(t_0 - u)}{\Omega(t_0 + k\pi / \Omega - u)} \left| \frac{\sin \Omega(t_0 - u)}{\Omega(t_0 - u)} \right| du = \pm\infty.$$

So the error after the filtering becomes ∞.

Remark: This is only an example for analysis. In the section of numerical results we will show that the bandpass filter (4) is not very effective for white noise.

THE REGULARIZED FILTERING ALGORITHM

First, we consider the regularized Fourier transform [6]:

$$\hat{f}_\alpha(\omega) = \int_{-\infty}^{\infty} \frac{f(t) e^{i\omega t} dt}{1 + 2\pi\alpha + 2\pi\alpha t^2},$$

where $\alpha > 0$ is the regularization parameter. Here $\hat{f}_\alpha(\omega)$ is the minimizer of a smoothing functional. We have proved $\hat{f}_\alpha(\omega)$ converges to the exact Fourier transform as the error of $f(t)$ approaches to zero. In [6], we have successfully used the regularized Fourier transform in extrapolation. So the weight function

$$\frac{1}{1 + 2\pi\alpha + 2\pi\alpha t^2}$$

is helpful to solve ill-posed problems.

Based on the regularized Fourier transform we present the regularized filtering formula:

$$f_\alpha(t) = \int_{-\infty}^{\infty} \frac{f(u)}{1 + 2\pi\alpha + 2\pi\alpha u^2} \frac{\sin \Omega(t - u)}{\pi(t - u)} du, \text{ a.e. } t \in \mathbf{R}. \tag{5}$$

where $f(t)$ is given in (2). The convergence property of this regularized filtering formula is given in the theorem below.

Theorem 3.1: For $f_E \in L^1$, if $\alpha \to 0$ and $\delta / \sqrt{\alpha} \to 0$ as $\delta \to 0$, then $f_\alpha \to f_E$ according to the maximum norm as $\delta \to 0$.

Proof:

$$\hat{f}_\alpha(t) - f_E(t) = -\int_{-\infty}^{\infty} \frac{f_E(u)(2\pi\alpha + 2\pi\alpha u^2)}{1+2\pi\alpha+2\pi\alpha u^2} \frac{\sin\Omega(t-u)}{\pi(t-u)} du + \int_{-\infty}^{\infty} \frac{\eta(u)}{1+2\pi\alpha+2\pi\alpha u^2} \frac{\sin\Omega(t-u)}{\pi(t-u)} du$$

where

$$\left| \int_{-\infty}^{\infty} \frac{\eta(u)}{1+2\pi\alpha+2\pi\alpha u^2} \frac{\sin\Omega(t-u)}{\pi(t-u)} du \right| \le \delta \int_{-\infty}^{\infty} \frac{1}{1+2\pi\alpha+2\pi\alpha u^2} du = \frac{\pi\delta}{\sqrt{2\pi\alpha(1+2\pi\alpha)}}.$$

For each $\varepsilon > 0$, there exists $T > 0$ such that $\int_{|u|>T} | f(u) | \, du < \varepsilon.$

Then

$$\left| \int_{-\infty}^{\infty} \frac{f_E(u)(2\pi\alpha + 2\pi\alpha u^2)}{1+2\pi\alpha+2\pi\alpha u^2} \frac{\sin\Omega(t-u)}{\pi(t-u)} du \right| \le \int_{-T}^{T} \frac{| f_E(u) | (2\pi\alpha + 2\pi\alpha u^2)}{1+2\pi\alpha+2\pi\alpha u^2} du + \int_{|u|>T} | f_E(u) | \, du$$

where

$$\int_{-T}^{T} \frac{| f_E(u) | (2\pi\alpha + 2\pi\alpha u^2)}{1+2\pi\alpha+2\pi\alpha u^2} du \to 0$$

as $\alpha \to 0$.

NUMERICAL RESULTS

In this section, we give some examples to show that the regularized filtering algorithm (5) is more effective in reducing the noise than the convolution (4). Suppose the exact signal in example 1 and 2 is

$$f_E(t) = \frac{1 - cost}{\pi t^2}.$$

Then

$$\hat{f}_E(\omega) = \begin{cases} 1 - |\omega|, & \omega \in [-\Omega, \Omega] \\ 0, & \omega \notin [-\Omega, \Omega] \end{cases}$$

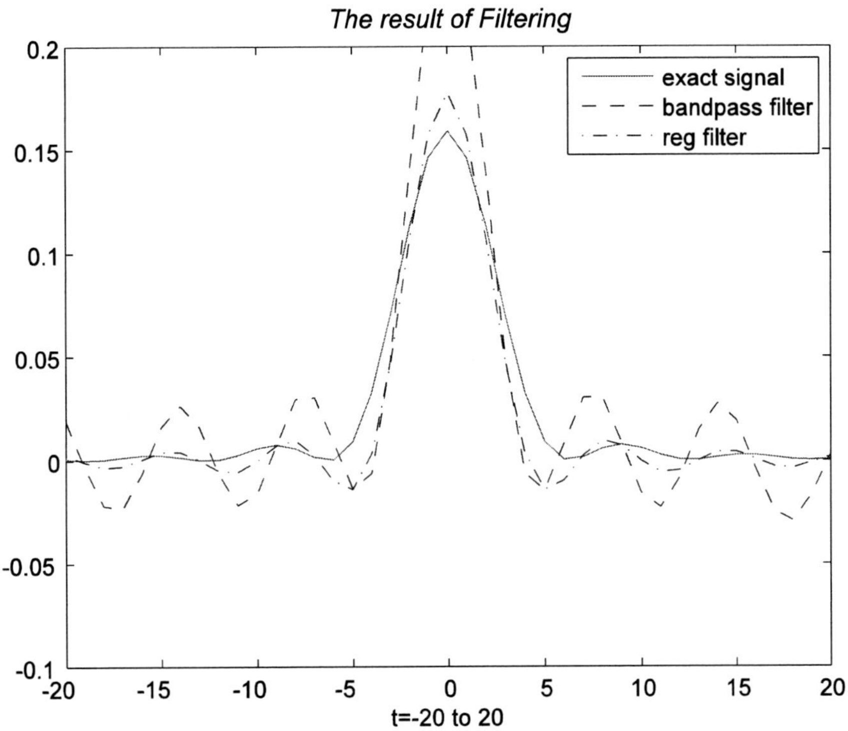

Figure 1. The numerical results of Example 1.

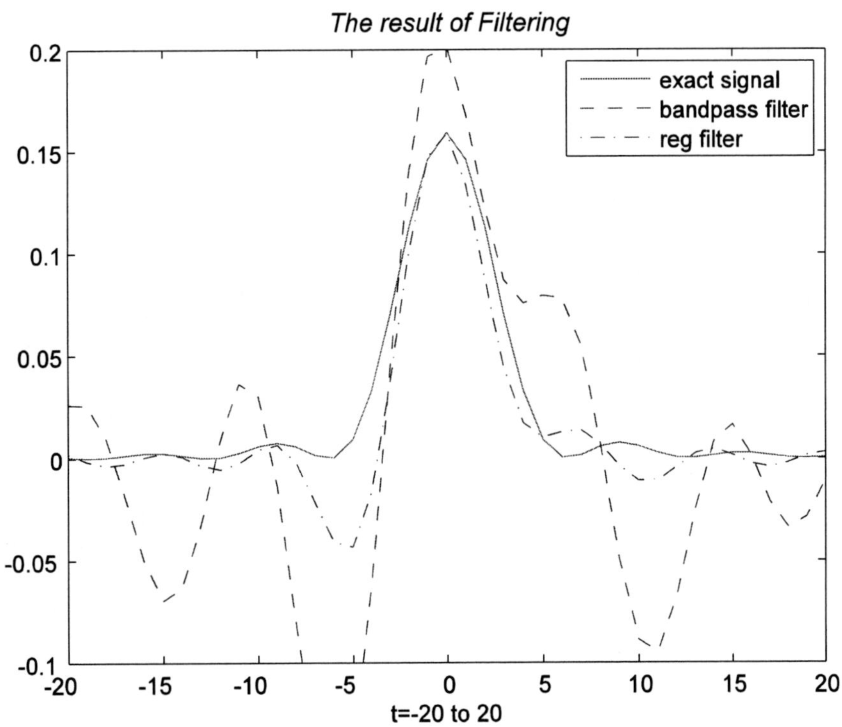

Figure 2. The numerical results of Example 2.

where $\Omega = 1$.

Example 1. We consider the noise

$$\eta(nh) = \varepsilon\ sgn\{\sin\Omega(t_0 - nh) / \Omega(t_0 - nh)\}$$

where $h = \pi / \Omega = \pi$, $t_0 = 0$, and $\varepsilon = 0.05$. This noise is used in the analysis of the stability in section 2. The result of (4) and the result of the regularized filtering algorithm with $\alpha = 0.02$ are in Figure 1.

Example 2. We consider the noise to be white noise that is Gauss distribution whose variance is 0.01.

The result of (4) and the result of the regularized sampling algorithm with $\alpha = 0.02$ are in Figure 2.

Conclusion

The filter of convolution with sinc function is not stable. For some noise the results of the filtering is not reliable. Regularized filtering algorithm is more effective in reducing the noise.

References

[1] D. Georgakopoulos and W. Q. Yang, "Circuit noise reduction by analogue lowpass filtering and data averaging", ELECTRONICS LETTERS 13th September 2001 Vol. 37 No. 19, pp. 1147- 1148.

[2] P. Daripa, W. Hua, "A numerical study of an ill-posed Boussinesq equation arising in water waves and nonlinear lattices: Filtering and regularization techniques", Elsevier Applied Mathematics and Computation 101 (1999), 159-207.

[3] O. Scherzer and J. Weickert, "Relations Between Regularization and Diffusion Filtering", *Journal of Mathematical Imaging and Vision*, 12, 43?3 (2000).

[4] T. Daboczi and I. Kollarl, "Multiparameter optimization of inverse filtering algorithms", IEEE Transactions On Instrumentation And Measurement, VOL. 45, NO. 2, APRIL 1996, pp. 417-421.

[5] D. Tschumperle and R. Deriche, "Vector-Valued Image Regularization with PDEs: A Common Framework for Different Applications", IEEE Transactions On Pattern Analysis And Machine Intelligence, VOL. 27, NO. 4, APRIL 2005 pp. 506-517.

[6] W. Chen,"An Efficient Method for An Ill-posed Problem—Band-limited Extrapolation by Regularization," IEEE Trans. on Signal Processing, vol. 54, pp.4611-4618, 2006.

[7] W. Chen, Band-Limited Extrapolation by Regularization, LAP Lambert Academic Publishing, 2010.

In: Digital Signal Processing (DSP)
Editor: Juan Zhang

ISBN: 978-1-63485-168-8

Chapter 5

DESIGN ASPECTS OF α-β-γ TRACKING FILTERS

Kenshi Saho*
Department of Electronic and Computer Engineering,
Ritsumeikan University, Kusatsu, Shiga, Japan

Abstract

This chapter discusses the design aspects of α-β-γ filters which are the simplest and widely-used digital tracking filters. Initially, the basics of a α-β-γ tracking filter are explained and its performance indices are introduced. The properties of each performance index are explained and suitable indices for appropriate design of a α-β-γ filter clarified. The design approaches for the filter gains are then introduced and compared. Primarily, a Kalman filter based (KF-based) and a minimum-variance (MV) filter approaches are reviewed. Although the KF-based approach is popular and well-used, there are cases where the appropriate gain setting is not obtained because of difficulties in the selection of the design parameter. In contrast, the gain design using the MV filter approach is easier than the KF-based approach. Numerical analyses and simulations show the effectiveness of the MV filter approach for practical use and indicate the problems of the conventional design approaches. Furthermore, the design of various types of α-β-γ filters including velocity measurements is also discussed. It is established that the KF-based approach design of the position-velocity-measured (PVM) α-β-γ filter is inappropriate because it is not derived from the general KF. The MV filter approach for α-β-γ filters using the position and velocity measurements is explained. Finally, numerical analyses show that the velocity-measured α-β-γ filters with MV filter criterion achieves accurate tracking compared with the general position-only-measured filters.

Keywords: moving object tracker, digital tracking filter design, α-β-γ filter, steady-state analysis, position and velocity measurements

*E-mail address: saho@fc.ritsumei.ac.jp

Introduction

The digital tracking filter outputs the predicted and smoothed (estimated) position, velocity, and acceleration of moving objects using the input parameters measured by remote monitoring systems. This technique is widely used in air traffic control, robots, vehicles, and security systems using various remote sensors such as radar, sonar, and optical cameras. Commonly used tracking filters are classified as the following two types [1]:

- Fixed-gain filter to realize a simple and low-cost system.

- Adaptive filter to achieve accurate estimation.

Kalman filters (KF), extended KF, and particle filters are known as adaptive tracking filters [2, 3, 4, 5, 6]. However, these techniques have a relatively heavy computational load, and in some cases they are impractical to use. Moreover, their design is conducted empirically, because it is difficult to evaluate the validity of the design parameters (e.g. the process noise) [1, 7]. In contrast, fixed-gain filters do not suffer from these problems. Examples of such filters are the α-β and α-β-γ filters [8, 9, 10, 11, 12, 13]. Because of their small computational load, they have been employed in various tracking systems such as motor position control [14], detecting falls [15], wireless sensor networks [16], and various other sensing applications [17, 18]. Moreover, there are only two or three gains, for which the performance indices can be analytically calculated. Consequently, it is simpler to design an appropriate filter than to construct adaptive tracking filters. However, although there are many determination algorithms for the gains and filter performance evaluation methods, their designs have barely been compared. In 2012, Ekstrand reviewed the design of tracking filters, including adaptive filters, and their problems [1]. However, he discussed only the fundamental properties of the α-β filter, and his paper did not sufficiently cover the design approaches for fixed-gain filters.

This chapter focuses on the design approaches for the gain of an α-β-γ filter (Note that the α-β filter is designed similarly to the α-β-γ filter and its formulations are relatively simple). First, the definitions and procedure for the α-β-γ filtering are explained. Filter performance indices to design appropriate tracking filter are then introduced. The properties and features of each index are explained. Two design approaches for the filter gains are reviewed: the Kalman filter based (KF-based) [13] and minimum-variance (MV) filter approaches [19]. Although the KF-based approach is known as an effective gain setting method, its problems are clarified in this chapter. Conversely, the MV filter approach, originally proposed by Kosuge and Kameda [20], is another effective approach. However, this is not well known because their paper is written in Japanese. Thus, this chapter explains its basics and effectiveness, and provides a comparison with the KF-based approach. Additionally, although the traditional α-β-γ filters use only position measurements, the position-velocity-measured (PVM) α-β-γ filters have recently been proposed and analyzed in[21, 22, 23, 24]. These have been applied to novel sensing systems such as micro-Doppler radar networks that can measure both position and velocity [25, 26, 27]. Thus, the PVM α-β-γ filters are widely applied and their design is important. The comparison of the design and achieved performance of the PVM and conventional α-β-γ filters are presented.

α-β-γ Filter Tracking

Definition

The α-β-γ *filter* predicts the position, velocity, and acceleration of a moving target, based on a constant acceleration model, using three filter gains [8, 9]. This is a one-dimensional tracking filter and is constructed in each axis for two (in xy plane) or three (in xyz space) dimensional tracking systems [7, 11]. This filter iterates prediction and smoothing processes. The prediction process is expressed by the following equations:

$$x_{\mathrm{p}k} = x_{\mathrm{s}k-1} + T v_{\mathrm{s}k-1} + (T^2/2) a_{\mathrm{s}k-1}, \tag{1}$$

$$v_{\mathrm{p}k} = v_{\mathrm{s}k-1} + T a_{\mathrm{s}k-1}, \tag{2}$$

$$a_{\mathrm{p}k} = a_{\mathrm{s}k-1}, \tag{3}$$

where $x_{\mathrm{s}k}$ is the smoothed target position at time kT, T the sampling interval, $x_{\mathrm{p}k}$ the predicted target position, $v_{\mathrm{s}k}$ the smoothed target velocity, $v_{\mathrm{p}k}$ the predicted target velocity, $a_{\mathrm{s}k}$ the smoothed target acceleration, and $a_{\mathrm{p}k}$ the predicted target acceleration. The smoothing process is expressed as follows:

$$x_{\mathrm{s}k} = x_{\mathrm{p}k} + \alpha(x_{\mathrm{o}k} - x_{\mathrm{p}k}), \tag{4}$$

$$v_{\mathrm{s}k} = v_{\mathrm{p}k} + (\beta/T)(x_{\mathrm{o}k} - x_{\mathrm{p}k}), \tag{5}$$

$$a_{\mathrm{s}k} = a_{\mathrm{p}k} + (\gamma/T^2)(x_{\mathrm{o}k} - x_{\mathrm{p}k}), \tag{6}$$

where $x_{\mathrm{o}k}$ is the measured target position, and α, β, and γ are filter gains. The block diagram of the α-β-γ filter is shown in Figure 1 where this filter is an infinite impulse response (IIR) filter with a single input and three outputs. The transfer function is easily derived as [8]:

$$G(z) = \frac{X_{\mathrm{p}}(z)}{X_{\mathrm{o}}(z)} = \frac{(\alpha\beta + \gamma/2)z^2 + (-2\alpha - \beta + \gamma/2)z + \alpha}{z^3 + (\alpha + \beta + \gamma/2 - 3)z^2 + (-2\alpha - \beta + \gamma/2 + 3)z + \alpha - 1}, \tag{7}$$

where the capital letters mean z-transformed functions. As shown in this transfer function, the α-β-γ filter is a three-dimensional low-pass filter and its gain coefficients are α, β, and γ.

Relationship to Kalman Filter

Generally, the α-β-γ filter is known as the steady-state *Kalman filter* [1, 7, 12, 13]. The Kalman filter estimates the target state vector based on the dynamic/measurement model. The dynamic model is:

$$\boldsymbol{x}_{\mathrm{t}k} = \boldsymbol{\Phi}\boldsymbol{x}_{\mathrm{t}k-1} + \boldsymbol{w}_k, \tag{8}$$

where $\boldsymbol{x}_{\mathrm{t}k}$ is the true target state vector at time kT, T the sampling interval, $\boldsymbol{\Phi}$ a transition matrix from kT to $(k+1)T$, and $\boldsymbol{w}_k$ the process noise, which has the covariance matrix $\boldsymbol{Q}$. The measurements of the target are modeled as:

$$\boldsymbol{z}_k = \boldsymbol{H}\boldsymbol{x}_{\mathrm{t}k} + \boldsymbol{v}_k, \tag{9}$$

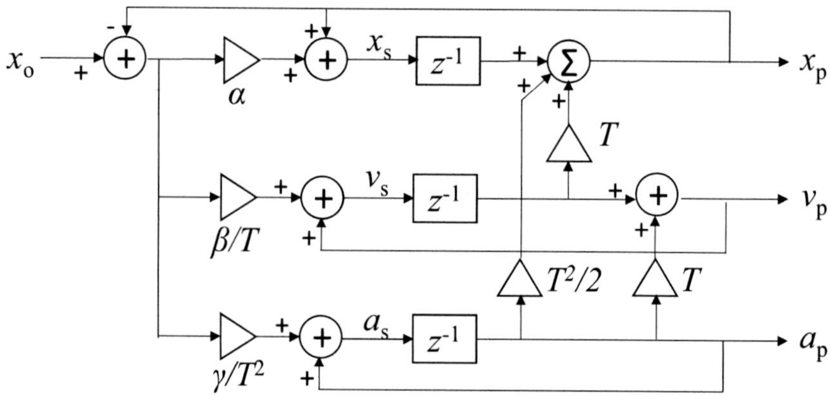

Figure 1. Block diagram of the α-β-γ filter.

where $\boldsymbol{z}_k$ is the measurement vector, $\boldsymbol{H}$ the measurement matrix, and $\boldsymbol{v}_k$ the measurement noise, which has the covariance matrix $\boldsymbol{R}$. The Kalman filter tracker for the above model sequentially estimates the state vectors via the following equations:

$$\tilde{\boldsymbol{x}}_k = \boldsymbol{\Phi}\hat{\boldsymbol{x}}_{k-1}, \tag{10}$$
$$\hat{\boldsymbol{x}}_k = \tilde{\boldsymbol{x}}_k + \boldsymbol{K}_k(\boldsymbol{z}_k - \boldsymbol{H}\tilde{\boldsymbol{x}}_k), \tag{11}$$

where forecasts and estimates are denoted by $\tilde{}$ and $\hat{}$, respectively, and $\boldsymbol{K}_k$ is the Kalman gain that minimizes the errors in the estimated state vectors. which is calculated as

$$\boldsymbol{K}_k = \tilde{\boldsymbol{P}}_k\boldsymbol{H}^{\mathrm{T}}(\boldsymbol{H}\tilde{\boldsymbol{P}}_k\boldsymbol{H}^{\mathrm{T}} + \boldsymbol{R})^{-1}, \tag{12}$$

where ^{T} denotes the transpose, and $\boldsymbol{P}_k$ is a covariance matrix of errors expressed as

$$\tilde{\boldsymbol{P}}_k = \boldsymbol{\Phi}\hat{\boldsymbol{P}}_{k-1}\boldsymbol{\Phi}^{\mathrm{T}} + \boldsymbol{Q}, \tag{13}$$
$$\hat{\boldsymbol{P}}_k = \tilde{\boldsymbol{P}}_k - \boldsymbol{K}_k\boldsymbol{H}\tilde{\boldsymbol{P}}_k. \tag{14}$$

Equations (10)–(14) are known as the Kalman filter equations.

The α-β-γ filter can be obtained from the Kalman filter equations by substituting following into (10)–(14):

$$\tilde{\boldsymbol{x}}_k = (x_{\mathrm{p}k} \quad v_{\mathrm{p}k} \quad a_{\mathrm{p}k})^{\mathrm{T}}, \tag{15}$$
$$\hat{\boldsymbol{x}}_k = (x_{\mathrm{s}k} \quad v_{\mathrm{s}k} \quad a_{\mathrm{s}k})^{\mathrm{T}}, \tag{16}$$
$$\boldsymbol{\Phi} = \begin{pmatrix} 1 & T & T^2/2 \\ 0 & 1 & T \\ 0 & 0 & 1 \end{pmatrix}, \tag{17}$$
$$\boldsymbol{H} = \begin{pmatrix} 1 & 0 & 0 \end{pmatrix}, \tag{18}$$
$$\boldsymbol{R} = (B_{\mathrm{x}}). \tag{19}$$

where B_{x} is the variance of errors in x_{o}. With $k \to \infty$, the steady-state Kalman gains are:

$$\boldsymbol{K}_\infty = \begin{pmatrix} \alpha \\ \beta/T \\ \gamma/T^2 \end{pmatrix}. \tag{20}$$

These equations mean that the steady-state gains are calculated for various settings of $\boldsymbol{Q}$ of (13). In other words, the α-β-γ filter obtained as the steady-state Kalman filter is limited to which can be represented by these equations. Thus, α-β-γ filters are not always the steady-state Kalman filter.

Filter Performance Indices

To evaluate the performance of the α-β-γ filters, various performance indices are used. For the tracking filter, tracking accuracy in steady-state is the most important. Thus, this section considers the indices of the steady-state errors. The indices explained in the following subsections are useful when designing α-β-γ filters.

$\tilde{\boldsymbol{P}}_\infty$ of the KF Equation

This is typically used to evaluate the tracking filter [28, 29]. The relationship between gains and $\tilde{\boldsymbol{P}}_\infty$ can be expressed analytically [12]. However, $\tilde{\boldsymbol{P}}$ overrates the variance in the errors caused by measurement noise, as verified by Ekstrand (see Section 9.8 of [1]). In addition, the relationship between basic properties such as the filter bandwidth and the error covariance matrix is not adequately clarified [1, 7]. Further, this is only usable for the α-β-γ filters that are obtained from the KF equations.

RMS Prediction Error

The RMS prediction error calculated using Monte Carlo simulations is an effective metric for the filter performance, because it represents a natural and practical index. Thus, most studies conduct Monte Carlo simulations to demonstrate improvements in the RMS prediction error using their proposed algorithms. However, for the filter design, an iterative trial-and-error process with varying filter parameters is required. Moreover, there are cases in which appropriate simulation settings are difficult to implement. Some researchers have used an analytically calculated RMS prediction error [30, 31], but its effectiveness is clarified for only a few limited tracking systems.

Measurement Error for Constant Velocity Target (Smoothing Performance Index)

An important function of the tracking filter is the reduction in random errors caused by measurement noise. One index for this performance is the steady-state error of a target under constant acceleration considering sensor noise. We assume the target moves with constant acceleration. The variance of the predicted target position in the steady-state is calculated using B_{x} and filter gains as [8, 9]:

$$\sigma_{\mathrm{p}}^2 = E[(x_{\mathrm{p}k} - x_{\mathrm{t}k})^2] = \frac{8\beta^2 + \alpha(4-2\alpha-\beta)(2\alpha\beta-\gamma(2-\alpha))}{(2-\alpha)(4-2\alpha-\beta)(2\alpha\beta-\gamma(2-\alpha))} B_{\mathrm{x}}, \tag{21}$$

where $x_{\mathrm{t}k}$ is the true target position and $E[\,]$ indicates the mean. Note that the mean error $E[x_{\mathrm{p}k} - x_{\mathrm{t}k}]$ is zero, because the assumed target motion is the same as the motion model of the α-β-γ filter (constant acceleration target). We call σ_{p}^2 the *smoothing performance index*.

Bias Error for Constant Acceleration Target (Tracking Performance Index)

The tracking filter is required to track complicated motion, including jerks. In the α-β-γ filter, steady-state bias error occurs when tracking a target moving with constant jerks, because the filter is based on a constant acceleration model. This error is an index of the tracking performance. When $x_{\mathrm{o}k} = J(kT)^3/6$ (J is the constant jerk) and the measurement errors are not considered, the steady-state predicted error is expressed as [10]:

$$e_{\mathrm{fin}} = \lim_{k\to\infty}(x_{\mathrm{o}k} - x_{\mathrm{p}k}) = JT^3/\gamma. \tag{22}$$

We call e_{fin} the *tracking performance index*. The smaller these tracking/smoothing performance indices, the better the tracking filter. However, the trade-off between e_{fin} and σ_{p}^2, is an important consideration in the design of tracking filters.

Filter Gain Design Approaches

KF-Based Approach

The main approach is to derive the gains for the α-β-γ filter from the steady-state Kalman filter [11, 12, 13]. Because the optimality in KF equations is guaranteed, this is quite a simple approach to obtaining appropriate gains. In this approach, the steady-state gains of the KF are calculated and are set as the gains of the α-β-γ filter. However, it is difficult to select appropriate process noise for the motion model. To overcome this difficulty, some typical and effective covariance matrices of the process noise and gain relationships obtained from them are known.

The most commonly used process noise is random acceleration, whose the covariance matrix is expressed as [13]:

$$\boldsymbol{Q}_{\mathrm{RA}} = \begin{pmatrix} T^4/4 & T^3/2 & T^2/2 \\ T^3/2 & T^2 & T \\ T^2/2 & T & 1 \end{pmatrix} \sigma_{\mathrm{aw}}^2. \tag{23}$$

In this model, $\boldsymbol{w}_k = (T^2/2 \quad T \quad 1)^{\mathrm{T}} w_{\mathrm{a}}$, where w_{a} is white Gaussian acceleration whose variance is σ_{aw}^2. This chapter calls the model with $\boldsymbol{Q}_{\mathrm{RA}}$ a random acceleration (RA) model, and also calls the α-β-γ filter derived from the RA model a RA filter. Another representative process noise is a random jerk process noise [12] and its $\boldsymbol{Q}$ is expressed as:

$$\boldsymbol{Q}_{\mathrm{RJ}} = \begin{pmatrix} T^6/36 & T^5/12 & T^4/6 \\ T^5/12 & T^4/4 & T^3/2 \\ T^4/6 & T^3/2 & T^2 \end{pmatrix} \sigma_{\mathrm{jw}}^2. \tag{24}$$

In this model, $\boldsymbol{w}_k = (T^3/6 \quad T^2/2 \quad T)^{\mathrm{T}} w_{\mathrm{j}}$, where w_{j} is a white Gaussian jerk whose variance is σ_{jw}^2. This chapter calls the model with $\boldsymbol{Q}_{\mathrm{RJ}}$ a random jerk (RJ) model, and also calls the α-β-γ filter derived from the RJ model a RJ filter.

For the above process noises, optimal gain relationships are derived and well-used in real applications. However, setting the variance of process noise ($\sigma_{\rm aw}^2$ or $\sigma_{\rm jw}^2$) is important in the filter design process because it directly determines performance. The useful parameter for this purpose is a tracking index proposed by Kalata [13]. This index is determined based on the degree of target maneuvering; if the target maneuvering motion is predicted as relatively large, λ is set to a large value. For the RA model, this is defined as:

$$\lambda = \sigma_{\rm aw} T^2 / \sqrt{B_{\rm x}}, \tag{25}$$

and $\sigma_{\rm jw}$ is used for the RJ model instead of $\sigma_{\rm aw}$. Thus the relationships between optimal gains and λ are derived [13, 32]. The filter design is realized by setting only λ and is easily conducted empirically. The relationships between the tracking index and the filter performance were investigated for some limited models such as the RA and RJ models. Crouse [33] recently derived the general solution for the optimal gains of arbitrary settings of the dynamic/measurement model for the KF-based approach. Using the above models and this solution, we can set the optimal gains for the assumed target model.

However, the α-β-γ filter whose gains are not derived from the KF is not designed from the KF-based approach. In addition, the performance of an α-β-γ filter derived from the Kalman filter is not optimal when evaluated using the performance indices expressed in (7) and (8) [19].

MV Filter Approach

To avoid the problems of the KF-based approach, the *MV filter* approach was proposed [10, 19, 20]. This approach determines the gains by minimizing the smoothing performance index $\sigma_{\rm p}^2$ under the condition that the tracking performance index $e_{\rm fin}$ is constant [19]. As shown in (22), the tracking performance index depends only on γ. Thus, for the general α-β-γ filter, the optimal gains with the MV filter approach are determined by:

$$\begin{aligned} &\arg\min_{\alpha,\beta}\ \sigma_{\rm p}^2 \\ &\text{sub. to}\quad \gamma = \text{const.} \end{aligned} \tag{26}$$

The solution for this problem satisfies [10]:

$$4\alpha\beta = \gamma(8 - 4\alpha - \beta), \tag{27}$$

$$\beta^3 + 2(3\alpha - 8)\beta^2 + 4(3\alpha^2 - 12\alpha + 16)\beta - 8\alpha^2(2 - \alpha) = 0. \tag{28}$$

The α-β-γ filter obtained from the above process is called a MV filter. As shown in the above process, the MV filter based approach, unlike the Kalman filter-based approach, does not require the process noise of the motion model [11, 12, 13]. The important finding of [19] is that there are some α-β-γ filters that are not derived from the KF equations. This is because the following assumptions for the KF are not considered in the MV filter approach:

- White Gaussian process noise is assumed.
- Motion is modeled as (nearly) constant acceleration (Constant jerk motion assumed in the tracking performance index is not considered).

Thus, the MV filter approach extends the range of the gains for the α-β-γ filters.

In [10], it was reported that the performance (evaluated using the tracking/smoothing performance indices of (21) and (22)) was better than that of other α-β-γ filters derived from the Kalman filter equations. Additionally, an efficient gain design based on MV filter approach was conducted by Tenne and Singh [8]. However, the performance analyses for the MV filter approach are based only on the theoretical analysis assuming constant acceleration or jerk targets. Thus, the comparison of the above approaches assuming a realistic maneuvering target is conducted in the next section.

Other Approaches

Other well-considered approaches are adaptive gain calculations based on an interacting-multiple-model [34] using optimization methods such as the genetic algorithm [35, 36] and particle swarm optimization [37], [38]. These methods can achieve accurate tracking compared with non-adaptive approaches. However, the computational load is heavy and the parameters of the optimization algorithms are empirically set. Moreover, the effectiveness of interacting-multiple-models compared with the KF and other extended filters is not clear, and further study of these other approaches is required.

Discussion Using Numerical Simulation

Design and evaluation of the α-β-γ filters can be summarized as follows based on the above reviews:

- The error covariance matrix of the KF equations cannot be used to evaluate the MV filter because this filter does not use the KF-based approach.
- The performance of the filters obtained from KF-based and MV filter approaches can be compared analytically because the tracking and smoothing performance indices use filter gains and variances of measurement noise.
- According to the analysis in [10], the MV filter achieves greater accuracy than the RA filter.

Thus, we cannot use the error covariance matrix to design arbitrary design parameters and the tracking/smoothing performance indices are effective. However, these indices only evaluate the steady-state accuracy of the filter, while other important properties such as convergence speed are not evaluated. The relationship between the realistic RMS errors and the performance indices is unclear. To use the performance indices for the appropriate filter design, the results of the analysis of the performance indices and the RMS errors obtained in the simulations must be close. This requires further investigation.

To evaluate for optimal filter performance, we first calculated the performance indices of the MV, RA, and RJ filters. As shown in (22), the tracking performance index depends only on γ. Thus, we calculated the smoothing performance index σ_{p}^2 for several γ with (21). We assumed T and B_{x} were normalized to 1. Table 1 shows the calculation results for γ=0.3 and 0.9. As shown in this table, the performances of the RA and RJ filters were almost the

Table 1. Analytical calculation results of σ_{p}^2 for γ=0.3 and 0.9

γ	MV filter	RA filter	RJ filter
0.3	2.40	3.23	3.23
0.9	6.23	11.8	11.2

same, and the MV filter achieved the best performance for both cases. For relatively large γ, the MV filter was effective and this was matched in the analysis in [9] and [10].

To compare the actual and analytical performance indices of Table 1, we used numerical simulations to investigate the performance of each filter as a maneuvering target in a two-dimensional (2D) tracking problem. Figure 2(a) shows the true trajectory of the target in the xy-plane, and Figure 2(b) and 2(c) show the true x and y at each point in time, respectively. Measurement parameters were obtained by adding white Gaussian noise to the true parameters. The sampling interval T was 10 ms, and the initial values were all zero. We compared the RMS errors given by the MV and RA filters and their calculation results for the smoothing performance index σ_{p}^2. Note that the RJ filter results were omitted because its performance was almost the same as the RA filter. The RMS error was calculated using 1000 Monte Carlo simulations which is defined as:

$$\epsilon_k = \frac{1}{1000}\sum_{i=1}^{1000}\sqrt{(x_{\mathrm{p}ik}-x_{\mathrm{t}k})^2+(y_{\mathrm{p}ik}-y_{\mathrm{t}k})^2}, \tag{29}$$

where $x_{\mathrm{p}ik}$ and $y_{\mathrm{p}ik}$ are $x_{\mathrm{p}k}$ and $y_{\mathrm{p}k}$ of $i-$th Monte Carlo simulation and $y_{\mathrm{t}k}$ is the true y. For 2D tracking, the α-β-γ filter is composed along each axis as follows:

$$\begin{pmatrix} x_{\mathrm{s}k} \\ v_{\mathrm{xs}k} \\ a_{\mathrm{xs}k} \\ y_{\mathrm{s}k} \\ v_{\mathrm{ys}k} \\ a_{\mathrm{ys}k} \end{pmatrix} = \begin{pmatrix} x_{\mathrm{p}k} \\ v_{\mathrm{xp}k} \\ a_{\mathrm{xp}k} \\ y_{\mathrm{p}k} \\ v_{\mathrm{yp}k} \\ a_{\mathrm{yp}k} \end{pmatrix} + \boldsymbol{K}\begin{pmatrix} x_{\mathrm{o}k}-x_{\mathrm{p}k} \\ y_{\mathrm{o}k}-y_{\mathrm{p}k} \end{pmatrix}, \tag{30}$$

$$\begin{pmatrix} x_{\mathrm{p}k} \\ v_{\mathrm{xp}k} \\ a_{\mathrm{xp}k} \\ y_{\mathrm{p}k} \\ v_{\mathrm{yp}k} \\ a_{\mathrm{yp}k} \end{pmatrix} = \begin{pmatrix} 1 & T & T^2/2 & 0 & 0 & 0 \\ 0 & 1 & T & 0 & 0 & 0 \\ 0 & 0 & 1 & 0 & 0 & 0 \\ 0 & 0 & 0 & 1 & T & T^2/2 \\ 0 & 0 & 0 & 0 & 1 & T \\ 0 & 0 & 0 & 0 & 0 & 1 \end{pmatrix}\begin{pmatrix} x_{\mathrm{s}k-1} \\ v_{\mathrm{xs}k-1} \\ a_{\mathrm{xs}k-1} \\ y_{\mathrm{s}k-1} \\ v_{\mathrm{ys}k-1} \\ a_{\mathrm{ys}k-1} \end{pmatrix}, \tag{31}$$

where $\boldsymbol{K}$ is the gain matrix, x_k, $v_{\mathrm{x}k}$, and $a_{\mathrm{x}k}$ denote position, velocity, and acceleration along the x-axis, y_k, $v_{\mathrm{y}k}$, and $a_{\mathrm{y}k}$ denote position, velocity, and acceleration along the y-axis, and subscripts "s", "p", and "o" denote "smoothed", "predicted", and "observed

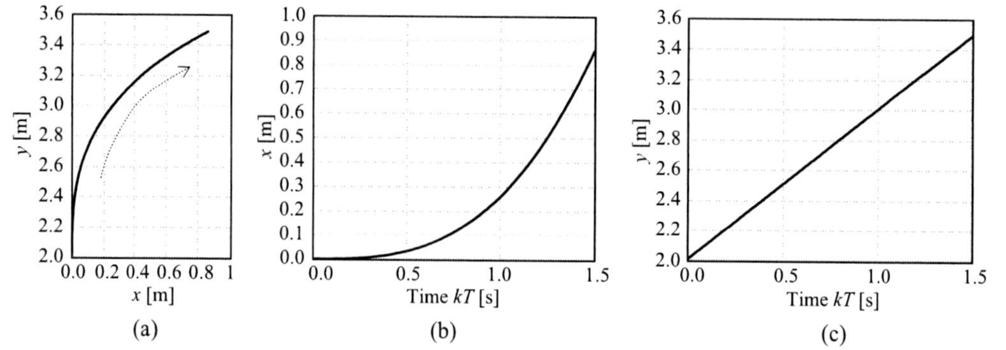

Figure 2. True motion of an assumed target in numerical simulation. (a) Target trajectory in xy, (b) Time variation of x, (c) Time variation of y.

(measured)", respectively. $\boldsymbol{K}$ is expressed as:

$$\boldsymbol{K} = \begin{pmatrix} \alpha & 0 \\ \beta/T & 0 \\ \gamma/T^2 & 0 \\ 0 & \alpha \\ 0 & \beta/T \\ 0 & \gamma/T^2 \end{pmatrix}. \tag{32}$$

Figure 3 shows the simulation results for $\gamma = 0.3$ (a) and 0.9 (b). In both cases, the MV filter achieved better steady-state accuracy than the RA filter. The difference in performance between these filters was relatively large for γ=0.9. These results were matched to Table 1 and indicated that the performance indices σ_{p}^2 and e_{fin} were useful for the design of filters including the MV filters, aside from the error covariance matrix. However, as shown in Figure 3, the convergence speed for the MV filter compared with the RA filter was worse in both cases. This indicates that the evaluation of the convergence speed is also important in real applications, although a performance index for this purpose was not considered. Consequently, comprehensive performance indices considering both accuracy and convergence speed are required and further research on this should be promising.

Position-Velocity-Measured α-β-γ Filter

The general α-β-γ filters reviewed above assumed a position measurement only because they are mainly used in the positioning systems for radar, sonar, and cameras. However, various sensing systems that can accurately measure velocity have recently been developed as a result of technical advances in various sensors and sensor networks, such as the micro-Doppler radar network [25, 26, 27]. Consequently, the application of tracking filters to such sensing systems has become an important area of research [39, 40, 41]. Although the number of studies is small, the *position-velocity-measured (PVM)* α-β-γ *filters* have been recently investigated [22, 23, 21, 24]. In this section, these are reviewed and their performance compared with the general α-β-γ filters. Moreover, some complementary analytical results that were omitted in [24] are also shown. Note

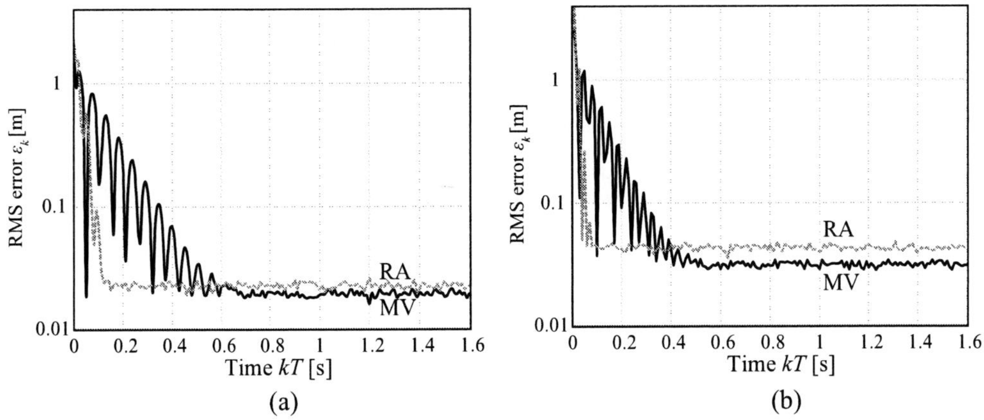

Figure 3. Simulation results for (a) γ=0.3 and (b) γ=0.9.

that the general MV filter described in previous sections is called the GMV filter in the following subsections.

Definition

We can define two types of PVM α-β-γ filters. The first has been used in several tracking systems that measure both position and velocity [22, 23]. However, its performance indices have not been derived, and thus the gain determination has so far been conducted empirically. The second type is a new PVM α-β-γ filter proposed in this chapter for the first time. The aim of this new filter is to achieve accurate tracking, even when the noise in the velocity measurements is comparatively large.

Using the measured velocity $v_{\mathrm{o}k}$, a PVM α-β-γ filter with the following smoothing process can be defined [22, 23, 21]:

$$x_{\mathrm{s}k} = x_{\mathrm{p}k} + \alpha(x_{\mathrm{o}k} - x_{\mathrm{p}k}), \tag{33}$$

$$v_{\mathrm{s}k} = v_{\mathrm{p}k} + \beta(v_{\mathrm{o}k} - v_{\mathrm{p}k}), \tag{34}$$

$$a_{\mathrm{s}k} = a_{\mathrm{p}k} + (\gamma/T)(v_{\mathrm{o}k} - v_{\mathrm{p}k}), \tag{35}$$

and the prediction process defined is the same as that in the general α-β-γ filter (expressed in (1)–(3)). Compared with the general α-β-γ filter, the second terms of (5) and (6) have been changed to use the measured velocity. Equations (34) and (35) show that the smoothed velocity and acceleration are estimated using the measured velocity. We call this PVM α-β-γ filter the Acceleration smoothed by measured Velocity (A-V) filter.

Because it uses the measured velocity, we expected the A-V filter to realize better tracking accuracy than the general α-β-γ filter. However, the performance of the A-V filter deteriorates when there is a large variance of measurement errors in velocity. To reduce this deterioration, we consider another PVM α-β-γ filter whose smoothing process is expressed as follows:

$$x_{\mathrm{s}k} = x_{\mathrm{p}k} + \alpha(x_{\mathrm{o}k} - x_{\mathrm{p}k}), \tag{36}$$

$$v_{\mathrm{s}k} = v_{\mathrm{p}k} + \beta(v_{\mathrm{o}k} - v_{\mathrm{p}k}), \tag{37}$$

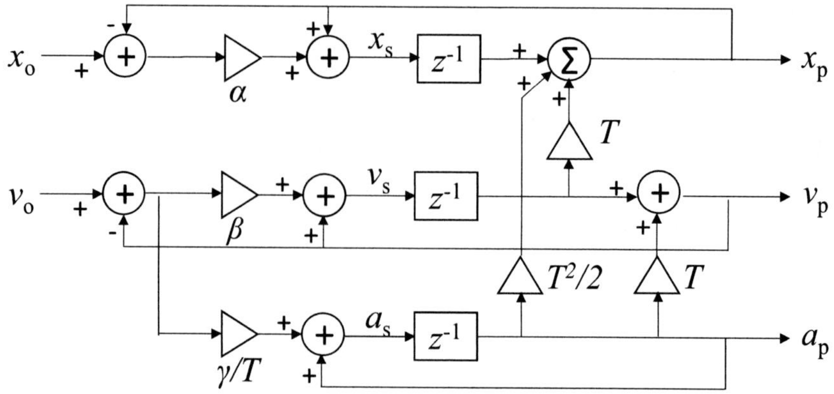

Figure 4. Block diagram of the A-V filter.

$$a_{sk} \quad = \quad a_{pk} + (\gamma/T^2)(x_{ok} - x_{pk}), \tag{38}$$

and whose prediction process is the same as in the general α-β-γ filter (i.e., (1)–(3)). The difference from the A-V filter is that the smoothed acceleration is estimated using the measured position, i.e., (6) in the general α-β-γ filter. We call this new PVM α-β-γ filter the Acceleration smoothed by measured Position (A-P) filter. It appears that the performance of the A-P filter is better than the A-V filter when the velocity noise is relatively large. In contrast, the A-V filter appears to outperform the A-P filter when the velocity noise is relatively small. Moreover, when the velocity is relatively large, it is unclear whether the performance of the A-P filter or the GMV filter is better.

Figures 4 and 5 show the block diagrams of the PVM filters. As indicated, these are two-input IIR filters aside from the conventional position-only-measured α-β-γ filters which have only a single input. The relationship between the inputs and the predicted position of the A-V filter is [24]:

$$X_p(z) = \frac{\alpha}{z+\alpha-1}X_o(z) + \frac{z((2\beta+\gamma)z - 2\beta+\gamma)}{2(z+\alpha-1)(z^2+(\beta+\gamma-2)z-\beta+1)}TV_o(z), \tag{39}$$

and that of the A-P filter is [24]:

$$X_p(z) = \frac{h_1(z)}{h_2(z)}X_o(z) + \frac{2\beta z(z-1)}{h_2(z)}TV_o(z), \tag{40}$$

where

$$h_1(z) \quad = \quad \gamma z^3 + (2\alpha - \beta\gamma + 2\gamma)z^2 + (2\alpha\beta - 4\alpha + \beta\gamma - 2\gamma)z - 2\alpha\beta - \beta\gamma + 2\alpha + \gamma, \tag{41}$$

$$\begin{aligned} h_2(z) \quad = \quad & (\gamma+2)z^3 + (2\alpha+2\beta+2\gamma-\beta\gamma-6)z^2 \\ & +(2\alpha\beta+\beta\gamma-4\alpha-4\beta-2\gamma+6)z - 2\alpha\beta - \beta\gamma + 2\alpha + 2\beta + \gamma - 2. \end{aligned} \tag{42}$$

As indicated, these are two-input low-pass filters.

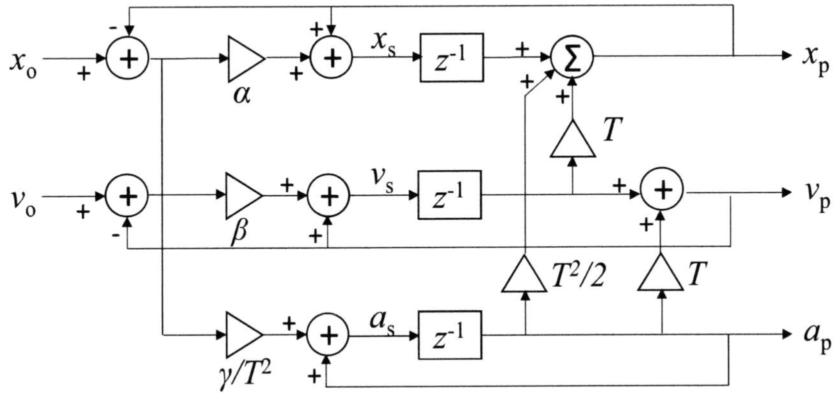

Figure 5. Block diagram of the A-P filter.

Relationship to KF

The relationship between the PVM α-β-γ filters and the KF is discussed here. We obtain the A-V filter by substituting the following into (10)–(14):

$$\tilde{\boldsymbol{x}}_k = (x_{\mathrm{p}k} \quad v_{\mathrm{p}k} \quad a_{\mathrm{p}k})^{\mathrm{T}}, \tag{43}$$

$$\hat{\boldsymbol{x}}_k = (x_{\mathrm{s}k} \quad v_{\mathrm{s}k} \quad a_{\mathrm{s}k})^{\mathrm{T}}, \tag{44}$$

$$\boldsymbol{\Phi} = \begin{pmatrix} 1 & T & T^2/2 \\ 0 & 1 & T \\ 0 & 0 & 1 \end{pmatrix}, \tag{45}$$

$$\boldsymbol{H} = \begin{pmatrix} 1 & 0 & 0 \\ 0 & 1 & 0 \end{pmatrix}, \tag{46}$$

$$\boldsymbol{R} = \begin{pmatrix} B_{\mathrm{x}} & 0 \\ 0 & B_{\mathrm{v}} \end{pmatrix}, \tag{47}$$

$$\boldsymbol{K}_{\mathbf{A-V}} = \begin{pmatrix} \alpha & 0 \\ 0 & \beta \\ 0 & \gamma/T \end{pmatrix}. \tag{48}$$

where B_{v} is the variance of errors in v_{o}. For the A-P filter, the gain matrix is:

$$\boldsymbol{K}_{\mathbf{A-P}} = \begin{pmatrix} \alpha & 0 \\ 0 & \beta \\ \gamma/T^2 & 0 \end{pmatrix}, \tag{49}$$

and the other matrices are the same as for the A-V filter. They indicate that the KF-based approach is unusable for designing the gains of the PVM filters because there are many zero components in $\boldsymbol{K}$. Thus, the tracking and smoothing performance indices and the MV filter approach are required for the A-V and A-P filters.

Performance Indices

To design the A-V and A-P filters similarly to the GMV filters with the MV filter approach, the tracking/smoothing performance indices are derived. Those of the A-V filter are [24, 21]:

$$\sigma^2_{\mathrm{p,A-V}} = \frac{\alpha}{2-\alpha}B_{\mathrm{x}} + \frac{f_1(\alpha,\beta,\gamma)}{f_2(\alpha,\beta,\gamma)}T^2 B_{\mathrm{v}}, \tag{50}$$

$$e_{\mathrm{fin,A-V}} = \frac{12-6\beta-\gamma}{12\alpha\gamma}JT^3, \tag{51}$$

where

$$\begin{aligned} f_1(\alpha,\beta,\gamma) &= \alpha^2(\beta-1)(4\beta^2-2\beta\gamma-\gamma^2+4\gamma) \\ &\quad +\alpha(6\beta^2\gamma-4\beta^3+8\beta^2+3\beta\gamma^2-16\beta\gamma-2\gamma^2+8\gamma) \\ &\quad -4\beta^2\gamma+2\beta\gamma(4-\gamma), \end{aligned} \tag{52}$$

$$f_2(\alpha,\beta,\gamma) = 2\alpha\beta(2-\alpha)(4-2\beta-\gamma)(\alpha^2+\alpha\beta+\gamma-\alpha^2\beta-\alpha\beta). \tag{53}$$

Additionally, the performance indices of the A-P filter are [24]:

$$\sigma^2_{\mathrm{p,A-P}} = \frac{g_1(\alpha,\beta,\gamma)B_{\mathrm{x}} + g_2(\alpha,\beta)T^2B_{\mathrm{v}}}{g_3(\alpha,\beta,\gamma)}, \tag{54}$$

$$e_{\mathrm{fin,A-P}} = JT^3/\gamma, \tag{55}$$

where

$$\begin{aligned} g_1(\alpha,\beta,\gamma) &= 8\alpha^3\beta(2-\beta)(\beta-1) \\ &\quad +2\alpha^2(\beta^3\gamma+4\beta^3-3\beta^2\gamma-8\beta^2+6\beta\gamma-4\gamma) \\ &\quad +\alpha\gamma(2\beta^3+\beta^2\gamma+4\beta^2-\beta\gamma-24\beta+16)-4\beta\gamma(2-\beta)^2, \end{aligned} \tag{56}$$

$$g_2(\alpha,\beta) = 8\beta^2(\alpha+\beta-\alpha\beta-2), \tag{57}$$

$$\begin{aligned} g_3(\alpha,\beta,\gamma) &= (16-8\beta-\beta\gamma-8\alpha+4\alpha\beta) \\ &\quad \cdot(2\alpha^2\beta^2-2\alpha^2\beta-2\alpha\beta^2+\alpha\beta\gamma-\alpha\gamma-2\beta\gamma+2\gamma), \end{aligned} \tag{58}$$

Gain Design Using the MV Filter Approach

The gain design of the A-V and A-P filters using the MV filter approach is presented in this section. First, the optimal gain calculation method for the A-V filter is presented. Under the MV filter criterion, we assume the tracking performance index is constant. With (51), the tracking performance index depends on:

$$C_{\mathrm{A-V}} = \frac{12-6\beta-\gamma}{12\alpha\gamma}. \tag{59}$$

Thus, $C_{\mathrm{A-V}}$ is constant in the MV filter criterion. Solving this for γ, we obtain:

$$\gamma = \frac{6(2-\beta)}{12\alpha C_{\mathrm{A-V}}+1}. \tag{60}$$

Substituting (60) into (50) gives the smoothing performance index $\sigma^2_{\mathrm{p,A-V}}(\alpha, \beta, C_{\mathrm{A-V}})$, which is used to calculate the optimal gains for constant $C_{\mathrm{A-V}}$. We then determine the optimal α and β for each $C_{\mathrm{A-V}}$ by:

$$\begin{aligned} &\arg\min_{\alpha,\beta}\ \sigma^2_{\mathrm{p,A-V}}(\alpha, \beta, C_{\mathrm{A-V}}) \\ &\text{sub. to}\quad C_{\mathrm{A-V}} = \text{const.} \end{aligned} \tag{61}$$

Next, we consider the optimal gain calculation for the A-P filter. As shown in (55), the tracking performance index of the A-P filter depends only on γ. Consequently, γ is constant when $e_{\mathrm{fin,A-P}}$ is constant. Thus, we determine the optimal α and β for each γ by:

$$\begin{aligned} &\arg\min_{\alpha,\beta}\ \sigma^2_{\mathrm{p,A-P}} \\ &\text{sub. to}\quad \gamma = \text{const.} \end{aligned} \tag{62}$$

Note that the above minimization problems are easily solved using the gradient descent method.

Performance Evaluation and Comparison with General Filters

To simplify the discussion, we define the following two parameters.

- The reciprocal of $C_{\mathrm{A-V}}$ is defined as:

$$\Gamma_{\mathrm{A-V}} = 1/C_{\mathrm{A-V}}. \tag{63}$$

  With (22), (51), and (55), $\Gamma_{\mathrm{A-V}}$ corresponds to γ in the A-P and GMV filters.

- The ratio of the two variances of measurement noise is defined as:

$$R_{\mathrm{v}} = T^2 B_{\mathrm{v}}/B_{\mathrm{x}}. \tag{64}$$

  The smoothing performance of the PVM α-β-γ filters depends on this ratio, as can be seen from (50) and (54). The relationship between R_{v} and the performance indices is important for the design of tracking filters that use both measured position and velocity.

The smoothing performance indices of the GMV, A-V, and A-P filters are calculated using (21), (50), and (54) under the assumption that the tracking performance indices are constant (i.e., $\Gamma_{\mathrm{A-V}}$ is constant for the A-V filter and γ is constant for the other filters). We assume that the sampling interval T and the variance of the measured position error B_{x} are normalized to 1.

Figure 6 shows the smoothing performance indices as a function of γ or $\Gamma_{\mathrm{A-V}}$ for various R_{v}. For relatively small R_{v}, shown in Figure 2(a), (b), and (c) ($R_{\mathrm{v}} = 1/2, 1,$ and 2), the PVM α-β-γ filters outperform the GMV filter, especially for large values of γ or $\Gamma_{\mathrm{A-V}}$. In these cases, the A-V filter realizes the best performance. This is because accurately measured velocities improve both smoothing and tracking performance. We can see from Figure 2(c) that the smoothing performance of the PVM α-β-γ filters is better than

Table 2. Summary of the properties of the MV α-β-γ filters considered in this chapter

	GMV filter	A-V filter	A-P filter
Measurement parameter	Position	Position and velocity	
Prediction process	(1)–(3)		
Smoothing process	(4)–(6)	(33)–(35)	(36)–(38)
Smoothing performance index	(21)	(50)	(54)
Tracking performance index	(22)	(51)	(22)
Suitable when R_v is	very large	small	large

that of the GMV filter when $0.2 \leq \gamma$ (or Γ_{A-V}) ≤ 1, even when $R_v = 2$. For larger values of R_v, shown in Figure 2(d), (e), and (f) ($R_v = 3, 5$, and 7), the performance deterioration in the proposed A-P filter is small compared with that in the A-V filter. This is because the smoothed acceleration in the A-P filter is calculated using the measured position. The smoothing performance of the A-P filter is better than that of the GMV filter when $\gamma \geq 0.4$ for $R_v = 5$ and $\gamma \geq 0.6$ for $R_v = 7$, as shown in Figure 2(e) and (f). These results imply that the proposed A-P filter can realize better performance than the GMV filter, even when the noise in the velocity measurements is considerable Figure 7 shows the smoothing performance indices as a function of R_v for γ and Γ_{A-V} values of 0.3, 0.6, and 0.9. For relatively small R_v, the A-V filter produces the best performance for all γ and Γ_{A-V}. When γ is large, the performance improvement using the measured velocity is also large. For relatively large R_v, the proposed A-P filter gives the best results. As shown in Figure 3(c), for $R_v = 10$ (i.e., the noise variance in the velocity measurements is 10 times as large as that in the position measurements), the A-P filter achieves better smoothing performance than the GMV filter. In contrast, the performance of the A-V filter deteriorates when R_v is comparatively large.

Table 2 summarizes the properties of the α-β-γ filters considered in this section. This table indicates that the A-V filter realizes accurate tracking for small R_v, whereas the proposed A-P filter realizes better accuracy than the other filters for relatively large R_v. Additionally, when R_v becomes large, the GMV filter realizes the best performance, which suggests that we should not use the measured velocity in this case. Based on these analyses, we can design the appropriate input parameters and filter gains for the PVM tracking systems.

Conclusion

This chapter discusses the design of various α-β-γ filters. From the viewpoint of the design aspects, the performance indices, design approaches, and PVM filters were reviewed, and some numerical simulations and analyses were conducted for discussion. The findings obtained by the reviews in this chapter are summarized below:

- Although the error covariance matrix in the KF equations is the well-used performance index, it is unsuitable because it overestimates the tracking errors and it is unusable for the filters without the use of the KF equations.

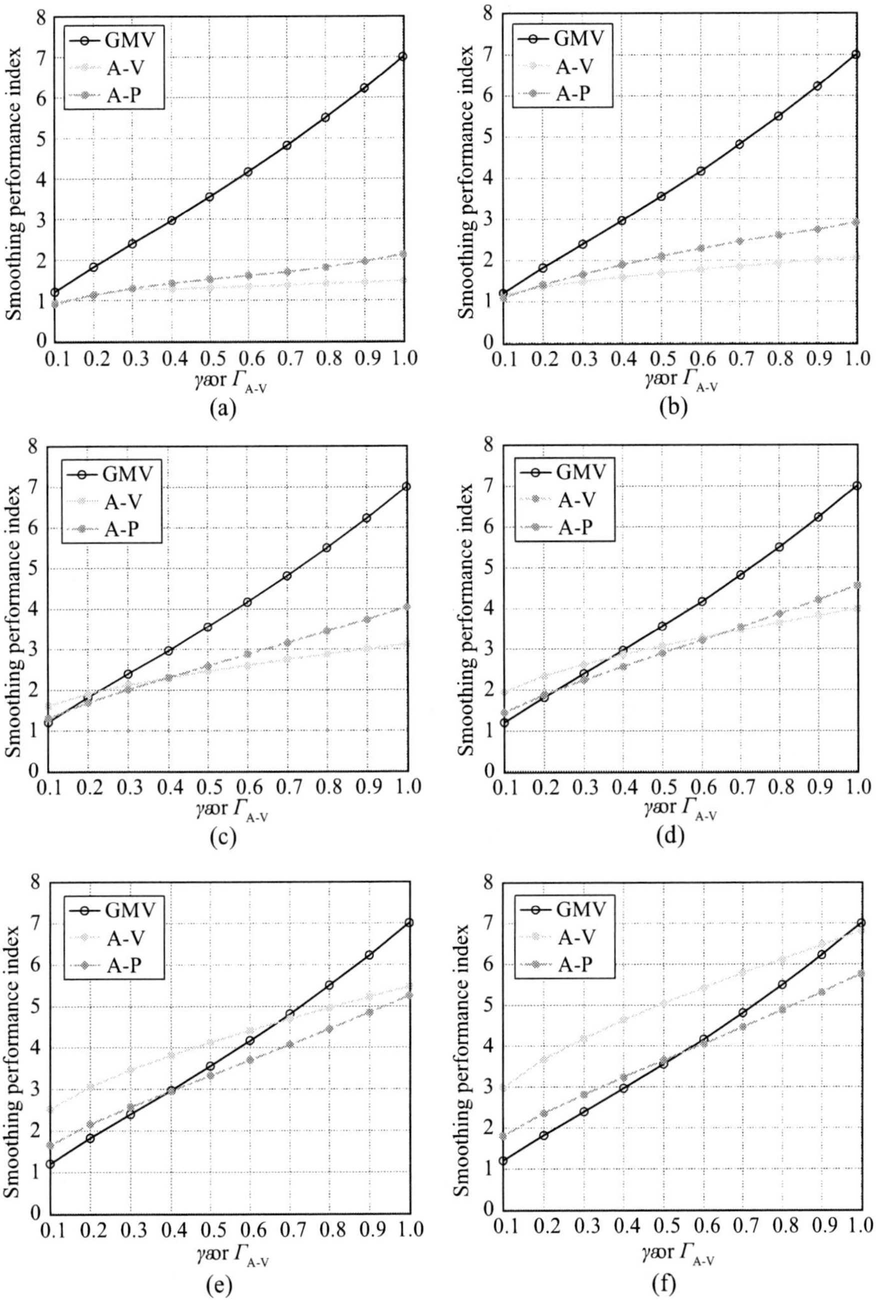

Figure 6. Relationship between smoothing performance indices and Γ_{A-V} or γ. (a) $R_{\mathrm{v}} = 1/2$, (b) $R_{\mathrm{v}} = 1$, (c) $R_{\mathrm{v}} = 2$, (d) $R_{\mathrm{v}} = 3$, (e) $R_{\mathrm{v}} = 5$, (f) $R_{\mathrm{v}} = 7$.

- The tracking and smoothing performance indices defined as (21) and (22) are effective for evaluating actual tracking errors. Analysis results using these indices almost match the RMS errors calculated using the Monte-Carlo simulations.

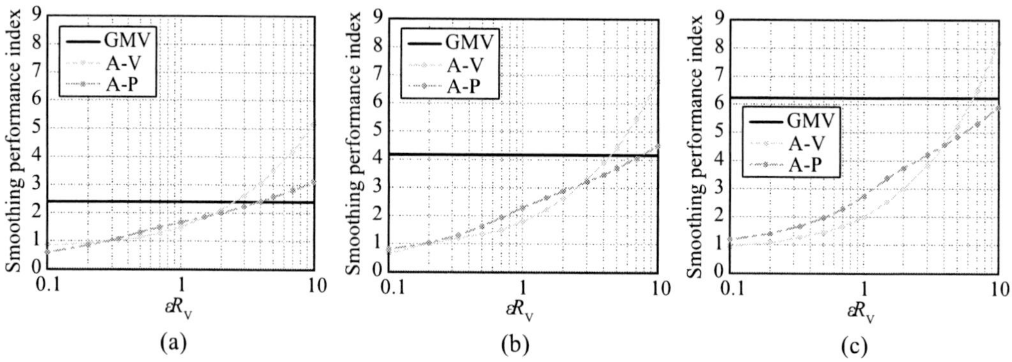

Figure 7. Relationship between smoothing performance indices and the ratio R_{v}. (a) $\Gamma_{A-V} = \gamma = 0.3$, (b) $\Gamma_{A-V} = \gamma = 0.6$, (c) $\Gamma_{A-V} = \gamma = 0.9$.

- The MV filter approach achieves accurate and simple designed gain based on the performance indices. The numerical simulation verifies that its steady-state accuracy is better than the RA filter which is derived from the KF-based approach.
- Two types of the PVM α-β-γ filters are defined: the A-V and A-P filters. These can improve the tracking accuracy of the general position-only-measured α-β-γ filters.
- The A-V and A-P filters are not derived from the KF-based approach. However, the MV filter approach is applicable for designing the gains of these filters and the effectiveness of the MV-PVM filters re clarified with numerical analyses.
- In the conventional performance indices, only accuracy is considered. The evaluation of other performance metrics such as convergence speed is promising future work.

References

[1] B. Ekstrand. Some aspects on filter design for target tracking. *J. Control Sci. Eng.*, page 870890, 2012.

[2] M. J. Jahromi and H. K. Bizaki. Target tracking in MIMO radar systems using velocity vector. *J. Inf. Sys. Telecommun.*, 2:150–158, 2014.

[3] K. Dae-Bong and H. Sun-Mog. Multiple-target tracking and track management for an FMCW radar network. *EURASIP J. Adv. Sig. Proc.*, 2013:159, 2013.

[4] H. Cheng, Z. Tao, and Z. Chao. Accurate three-dimensional tracking method in bistatic forward scatter radar. *EURASIP J. Adv. Sig. Proc.*, 2013:66, 2013.

[5] H. Niknejad, A. Takeuchi, S. Mita, and D. McAllester. On-road multivehicle tracking using deformable object model and particle filter with improved likelihood estimation. *IEEE Trans. on Intel. Transport. Sys.*, 13:748–758, 2012.

[6] M. Daun and F. Ehlers. Tracking algorithms for multistatic sonar systems. *EURASIP J. Adv. Sig. Proc.*, 2010:461538, 2010.

[7] Y. Bar-Shalom, X.R. Li, and T. Kirubarajan. *Estimation with Applications to Tracking and Navigation*. Wiley-Interscience, New York City, United States, 2001.

[8] D. Tenne and T. Singh. Characterizing performance of α-β-γ filters. *IEEE Trans. Aero. Elec. Sys.*, 38:1072–1087, 2002.

[9] Y. Kosuge, M. Ito, T. Okada, and S. Mano. Steady-state errors of an α-β-γ filter for radar tracking. *Electron. and Commun. in Japan (Part III: Fundamental Electronic Science)*, 85:65–79, 2002.

[10] Y. Kosuge and M. Ito. A necessary and sufficient condition for the stability of an α-β-γ filter. In *Proc. of the 40th SICE Annual Conf.*, pages 7–12, Nagoya, July 2001.

[11] C. C. Arcasoy and G. Ouyang. Analytical solution of α-β-γ tracking filter with a noisy jerk as correlated target maneuver model. *IEEE Trans. Aero. Elec. Sys.*, 33:347–353, 1997.

[12] J. J. Sudano. The α-β-γ tracking filter with a noisy jerk as the maneuver model. *IEEE Trans. Aero. Elec. Sys.*, 29:578–580, 1993.

[13] P. R. Kalata. The tracking index: A generalized parameter for α-β and α-β-γ target trackers. *IEEE Trans. Aero. Elec. Sys.*, AES-20:174–182, 1984.

[14] N. H. Khin, Y. F. Che, S. M. L. Eileen, and W. X. Liang. Alpha beta gamma filter for cascaded PID motor position control. *Procedia Engineering*, 41:244–250, 2012.

[15] Y. S. Lee and H. J. Lee. Multiple object tracking for fall detection in real-time surveillance system. In *Proc. of Int. Conf. Advanced Communication Technology 2009 (ICACT2009)*, pages 2308–2312, Phoenix Park, February 2009.

[16] S. Sharma, S. Deshpande, and K. M. Silvalingam. Alpha-beta filter based target tracking in clustered wireless sensor networks. In *Proc. of 3rd Int. Conf. Communication Systems and Networks*, pages 1–4, Bangalore, India, January 2011.

[17] Y. Wang. Feature point correspondence between consecutive frames based on genetic algorithm. *Int. J. of Robotics and Automation*, 21:35–38, 2006.

[18] K. Daniilidis, C. Krauss, M. Hansen, and G. Sommer. Real-time tracking of moving objects with an active camera. *Real-Time Imaging*, 4:3–20, 1998.

[19] Y. Kosuge and M. Ito. Evaluating an α-β filter in terms of increasing a track update-sampling rate and improving measurement accuracy. *Electron. Commun. in Japan (Part I: Communications)*, 86:10–20, 2003.

[20] Y. Kosuge and H. Kameda. Optimal gains of an - filter for radar tracking. *IEICE Trans. A (Japanese edition)*, J82-A:351–364, 1999.

[21] C. Wei, T. Jing, and W. SiLiang. A tracking loop based on tightly coupled range and velocity filter equations. *Science China-Information Sciences*, 55:898–910, 2012.

[22] H. Yamazaki, K. Saho, and T. Sato. Accurate shape estimation method for multiple moving targets with UWB Doppler radar interferometers. In *Proc. of 10th International Conference on Space, Aeronautical and Navigational Electronics*, pages 7–12, Hanoi, December 2013.

[23] C. Zheng. Tracking vehicular motion-position using V2V communication. Master's thesis, the University of Waterloo, 2010.

[24] K. Saho and M. Masugi. Performance analysis of alpha-beta-gamma tracking filters using position and velocity measurements. *EURASIP J. Adv. Sig. Process.*, 2015:35, 2015.

[25] K. Saho, T. Sakamoto, T. Sato, K. Inoue, and T. Fukuda. Pedestrian imaging using UWB Doppler radar interferometry. *IEICE Trans Commun.*, E96-B:613–623, 2013.

[26] R. Kozma, L. Wang, K. Iftekharuddin, E. McCracken, M. Khan, K. Islam, S. R. Bhurtel, and R. M. Demirer. A radar-enabled collaborative sensor network integrating COTS technology for surveillance and tracking. *Sensors*, 12:1336–1351, 2012.

[27] J. H. Lim, A. Terzis, and I-J. Wang. Tracking a non-cooperative mobile target using low-power pulsed Doppler radars. In *Proc. of 2010 IEEE 35th Conf. Local Computer Networks (LCN)*, pages 913–920, Denver, CO, October 2010.

[28] J. H. Painter, D. Kerstetter, and S. Jowers. Reconciling steady-state kalman and alpha-beta filter design. *IEEE Trans. Aero. Elec. Sys.*, 26:986–991, 1990.

[29] L. Yujing, M. Huadong, and W. Desheng. teady-state analysis of target tracker with constant input/bias constraint. *Tshinghua Sci. Tech.*, 13:35–42, 2008.

[30] V. Jain and W. D. Blair. Filter design for steady-state tracking of maneuvering targets with lfm waveforms. *IEEE Trans. Aero. Elec. Sys.*, 45:765–773, 2009.

[31] T. L. Ogle and W. D. Blair. Fixed-lag alpha-beta filter for target trajectory smoothing. *IEEE Trans. Aero. Elec. Sys.*, 40:1417–1421, 2004.

[32] J. E. Gray and W. Murray. A derivation of an analytic expression for the tracking index for the alpha-beta-gamma filter. *IEEE Trans. Aero. Elec. Sys.*, 29:1064–1065, 1993.

[33] D. F. Crouse. A general solution to optimal fixed-gain (α-β-γ etc.) filters. *IEEE Sig. Process. Lett.*, 22:901–904, 2015.

[34] D. Mohammed, K. Mokhtar, O. Abdelaziz, and M. Abdelkrim. A new IMM algorithm using fixed coefficients filters (fastIMM). *Int. J. of Electron. Commun. (AEÜ)*, 64:1123–1127, 2009.

[35] C.M. Wu, C.K. Chang, and T.T. Chu. he optimiation of parameters for target tracking filter by genetic algorithm and Taguchi method. *Pakistan J. Statistics*, 28:167–176, 2012.

[36] W. Chun-Mu, C. Ching-Kao, and C. Tung-Te. A new EP-based α-β-γ-δ filter for target tracking. *Mathematics and Computers in Simulation*, 81:1785–1794, 2011.

[37] T. E. Lee, J.-P. Su, and K.-W. Yu. A particle swarm optimization-based state estimation scheme for moving objects. *Trans. Inst. Measur. Control*, 34:236–254, 2012.

[38] M. B. Basneta J. Mara and C. C. Liua. Sea vehicle's motion compensation using adaptive alpha-beta-gamma filter. *J. Chinese Inst. Eng.*, 3:277–285, 2015.

[39] Y. Ju Hong, K. Du Yong, B. Seung Hwan, and S. vladimir. Joint initialization and tracking of multiple moving objects using Doppler information. *IEEE Trans. Sig. Proc.*, 59:3447–3452, 2011.

[40] B.R. Geetha and K. V. Ramachandra. A three state Kalman filter with range and range-rate measurements. *Int. J. Computer Application*, 3:85–101, 2013.

[41] X. Zhu, J. Hong, and W. Cui. Study on radar data processing algorithm with improved Kalman filter. In *Proc. of 4th IEEE Conf. on Industrial Electronics and Applications 2009*, pages 3826–3829, Xi'an, May 2009.

In: Digital Signal Processing (DSP)
Editor: Juan Zhang

ISBN: 978-1-63485-168-8

Chapter 6

DIGITAL SIGNAL MODEL OF ONE-DIMENSIONAL PHOTONIC CRYSTAL

Juan Zhang* and Yang Wang†
[1]School of Communication and Information Engineering,
Shanghai University, Shanghai, China
[2]Shanghai Institute of Optics and Fine Mechanics,
Chinese Academy of Sciences, Shanghai, China

ABSTRACT

A method of describing one-dimensional photonic crystals (1DPCs) based on Z-domain digital signal processing theory is presented in this chapter. The analytical expression of the target band gap spectrum in the digital domain is obtained by the autocorrelation of its impulse response. The feasibility of this method is verified by reconstructing two simple 1DPC structures with a target photonic band gap obtained by the traditional transfer matrix method. This method provides an effective approach to function-guided designs of interference-based band gap structures for photonic applications.

Keywords: one-dimensional photonic crystal, digital signal processing, Z-transform

INTRODUCTION

Photonic crystal (PC) [1] is a periodic nanostructure composed of regularly repeating regions of high and low refractive indices that affect the propagation of light waves in the same way that ionic lattices affect electrons in solids. One-dimensional PC (1DPC), often in the form of periodic multilayer dielectric stacks, is simpler in design and fabrication and has already been extensively used in optics and photonics [2-7]. To design a PC system, the

---

* Juan Zhang e-mail: juanzhang@staff.shu.edu.cn.
† Yang Wang e-mail: ywang@siom.ac.cn.

location and width of the photonic band gap is essential to engineer. This engineering feat is often achieved by computational method such as plane wave expansion method, finite difference time domain method, transfer matrix method, etc. The band structure and output spectrum of a PC can be obtained if its structural parameters (such as refractive index and thickness of each layer in a 1DPC) are given. However, in contrast to this forward process, the optimized PC structure is challenging to optimize from a target spectrum or band structure, which is more important for the design of a functional PC system. Hence, effective function-guided PC design methods are urgently needed.

Digital signal processing (DSP) is the mathematical manipulation of a digitized discrete signal to modify or improve it in a specific way (such as filtering). In recent years, the rapid development of computer and microelectronics technology has enabled DSP technology to enter into a new stage and be extensively used in communications, audio/image processing, system control, etc. [8]. Based on the similar impulse response nature of optical interference, the DSP algorithm becomes a powerful tool for designing and analyzing optical filters [9-12]. The intrinsic characteristics of this computer-aided method make it more suitable for complex structures and function-guided designs.

In fact, some numerical techniques besides DSP method have been suggested and used to design thick optical thin-film filters [13], finite length fiber gratings [14], non-uniform fiber Bragg gratings [15] and other similar structures [16] previously. These techniques are based on the differential layer-peeling/stripping or inverse scattering algorithm [17, 18]. For this design method, although no knowledge of a reasonable starting guess at the final structure is necessary, both the phase and amplitude reflectance data are needed.

In this chapter, an effective method of designing a 1DPC filter based on the Z-domain DSP technique is presented. The target reflectance/transmittance spectrum with an optical stopband is described analytically by an accurate rational function and calculated out through its autocorrelation in the digital domain. And then the structural parameters of the 1DPC can be obtained by the polynomial recursion method (a layer-peeling algorithm). It should be pointed out that no phase information of the target spectrum is needed in our calculation, thus it is simpler than the previous methods. The optical parameters of the 1DPC are mathematically described from the DSP point of view and thereby become compatible with common MATLAB programs [19] and digital signal processors [20].

THEORY AND NUMERICAL METHODS

The optical multilayer dielectric film can be considered as a special case of 1DPC. The traditional multilayer dielectric mirrors show typical photonic band gap characteristics. An optical multilayer filter can be mathematically synthesized by Fourier transform or Z-transform [21], which is an important basis of DSP theory. As such, 1DPCs can be described by DSP theory through the multilayer filter. The transmission of light through a certain layer of a multilayer structure and its corresponding Z-domain block diagram are given in Figure 1 [21]; 0, 1, and 2 refer to the incident medium, the layer, and the exit medium, respectively. r_{01}, t_{01} and r_{12}, t_{12} are the reflection and transmission coefficients of light on the 0-to-1 and 1-to-2 surfaces, respectively. r_{10}, t_{10} are the reflection and transmission coefficients of light on the 1-to-0 surface. The reflection of light incident on the interface from the left and right sides is opposite to each other, i.e., $r_{01} = -r_{10}$. E_i, E_r, and E_t are electric fields of incident, reflected, and transmitted light, respectively. z^{-1} represents the time delay or path difference of the light after passing through the layer. $z^{-1} = e^{-j2\pi nfd/c} = e^{-j\omega nd/c}$, where c is the speed of light in vacuum, n is the refractive index of the layer, f is the frequency of light, d is the thickness of the layer, and + represents the superposition operation.

From the correspondence between the optical and DSP model shown in Figure 1, the transfer functions of transmitted and reflected light can be obtained as

$$H_t(z) = \frac{E_t(z)}{E_i(z)} = \frac{t_{01}t_{12}z^{-1}}{1-r_{10}r_{12}z^{-2}}, \tag{1}$$

$$H_r(z) = \frac{E_r(z)}{E_i(z)} = r_{01} + \frac{t_{01}t_{10}r_{12}z^{-2}}{1-r_{10}r_{12}z^{-2}}. \tag{2}$$

Both of them are in the form of typical rational function, and the numerator and denominator polynomials have the same order of z^{-1} for the reflection transfer function. The transfer functions of the multilayer structure can be calculated by cascading the Z-domain block diagrams and with similar rational function forms.

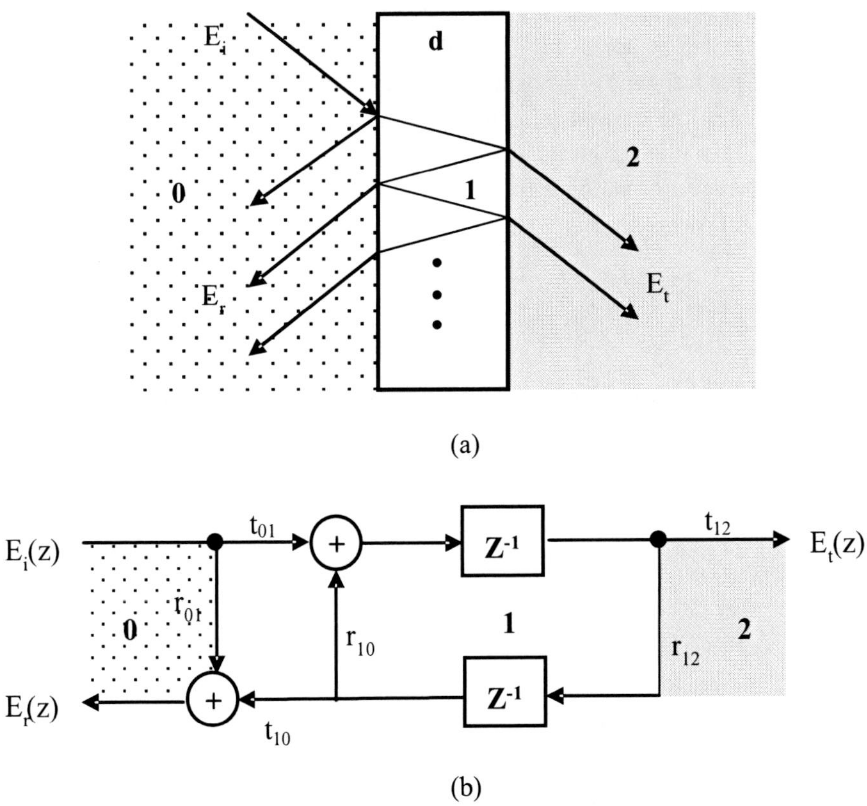

Figure 1. (a) Transmission of light through a single layer and (b) its Z-domain block diagram.

In the DSP theory, for a rational function in the following form (the numerator and denominator polynomials have the same order of z^{-1}):

$$H(z)=\frac{B(z)}{A(z)}=\frac{b_0+b_1z^{-1}+b_2z^{-2}\dots\dots\dots+b_pz^{-p}}{a_0+a_1z^{-1}+a_2z^{-2}\dots\dots+a_pz^{-p}}, \qquad (a_0=1) \tag{3}$$

the autocorrelation of its impulse response should satisfy the following relationship [5]:

$$r_{hh}(m)=\begin{cases} -\sum_{k=1}^{p}a_k r_{hh}(m-k), & m>p \\ -\sum_{k=1}^{p}a_k r_{hh}(m-k)+\sum_{k=0}^{p-m}h(k)b_{k+m}, & 0\le m\le p, \\ r^{*}_{hh}(m), & m<0 \end{cases} \tag{4}$$

where h(k) is the impulse response of H(z).

For the desired filtering spectrum, the autocorrelation of its impulse response can be easily obtained, so the coefficient a_k (k = 0, 1 … p) can be directly determined from the case of m > p in Equation (4). Thus, the denominator expression of the desired target reflection transfer function A(z) can be achieved. Then, the frequency response of B(z) can be obtained from the product of the frequency response of the target reflection spectrum and A(z). For the numerator expression of the desired reflection transfer function, its coefficient b_k (k = 0, 1 … p) cannot be directly obtained from its autocorrelation. The coefficient b_k (k = 0, 1 … p) can be obtained by constructing a rational function C(z), which has the same frequency response as B(z) and has the following form:

$$C(z)=1/(c_0+c_1z^{-1}+c_2z^{-2}+\dots\dots\dots\dots+c_qz^{-q}), \tag{5}$$

where q>>p. The form of Equation (5) is the same as that of Equation (3), in which b_0 = 1 and b_k = 0, so the coefficient c_k (k = 0, 1 … q) can be calculated from Equation (4) in a way similar to that of calculating a_k (k = 0, 1 … p). Given that B(z) and C(z) have the same frequency response in the Z-domain, then

$$B(z)=1\Big/\sum_{i=0}^{q}c_iz^{-i}. \tag{6}$$

Equation (6) shows that the coefficients of B(z) and C(z) are related by a convolution sum. Accordingly, we can construct the square error as

$$\xi=\sum_{k=0}^{p+q}\left[b_0c_k+\sum_{n=1}^{k}b_kc_{k-n}\right]^2-1. \tag{7}$$

The coefficients of B(z) can be obtained when (has the minimum value). Hence, b_k (k = 0, 1 … p) can be determined by the following relationship:

$$\begin{pmatrix} b_1 \\ b_2 \\ \vdots \\ b_p \end{pmatrix} = -\begin{pmatrix} r_{cc}(0) & r_{cc}(-1) & \cdots & r_{cc}(-p+1) \\ r_{cc}(1) & r_{cc}(0) & \cdots & r_{cc}(-p+2) \\ \vdots & \vdots & \vdots & \vdots \\ r_{cc}(p-1) & r_{cc}(p-2) & \cdots & r_{cc}(0) \end{pmatrix}^{-1} \begin{pmatrix} r_{cc}(1) \\ r_{cc}(2) \\ \vdots \\ r_{cc}(p) \end{pmatrix}, \tag{8}$$

where $r_{cc}(\tau)$ (τ = -p + 1, -p + 2, p) is the autocorrelation of coefficients c_k (k = 0, 1 … q), i.e.,

$$r_{cc}(\tau) = \sum_{k=0}^{q} c_k c^*_{k+\tau}, \tag{9}$$

where the superscript * indicates conjugation. Notably, all these coefficients should be normalized.

To date, all coefficients of the denominator and numerator of the target reflection transfer function have been obtained, and the expression of the target reflection spectrum in the digital domain with the order of p is known. We need to optimize the expression to obtain the optimum design result. The optimized expression can be obtained by increasing the order and constantly comparing it with the target reflection transfer function. Once the optimized expression of the target reflection spectrum in the digital domain is achieved, the reflection coefficients of each interface can be obtained by the recursion algorithm [21]. The reflection coefficient of an interface has a certain relationship with the refractive indexes of the two adjacent media [11]. Thus, the refractive index of each layer in the structure can be easily obtained from the reflection coefficients of each interface. The flow diagram for obtaining the optimized expression of the transfer function is shown in Figure 2. The optical thickness of the medium can be determined by the period of the target spectrum. The period of the target spectrum is related to the parameter z in the digital domain.

DESIGN EXAMPLE

The feasibility of the proposed design method is confirmed by using two simple examples. An 11-layer 1DPC with a band gap near 1.55 μm (fiber communication wavelength) is set as the design target. The solid curve in Figure 3 shows the transmittance spectrum of the $(HL)^5H$ structure obtained by the traditional transfer matrix method, where the low refractive index material (L) is magnesium fluoride with a refractive index of 1.38, and the high refractive index material (H) is zinc sulfide with a refractive index of 2.35 at 1.55 μm (λ). The optical thickness of each layer is equal to 1/4λ.

The reflection transfer function in the digital signal domain of the target spectrum shown in Figure 3 can be calculated as

$$H(z) = \frac{0.4132 - 0.5570z^{-1} + 0.7711z^{-2} - 0.9643z^{-3} + 1.1064z^{-4} - 1.1828z^{-5}}{1 - 0.8125z^{-1} + 1.0008z^{-2} - 1.1262z^{-3} + 1.1826z^{-4} - 1.1563z^{-5}}$$

$$\frac{+1.1758z^{-6} - 1.0944z^{-7} + 0.9424z^{-8} - 0.7499z^{-9} + 0.5341z^{-10} - 0.3946z^{-11}}{+1.0574z^{-6} - 0.8975z^{-7} + 0.7029z^{-8} - 0.4973z^{-9} + 0.3076z^{-10} - 0.1623z^{-11}}. \quad (10)$$

As shown in Equation (10) and according to the recursion algorithm, the reflection coefficients of each interface can be obtained, which are shown in Table 1. Then, the refractive indexes of each layer can be calculated as shown in Table 2.

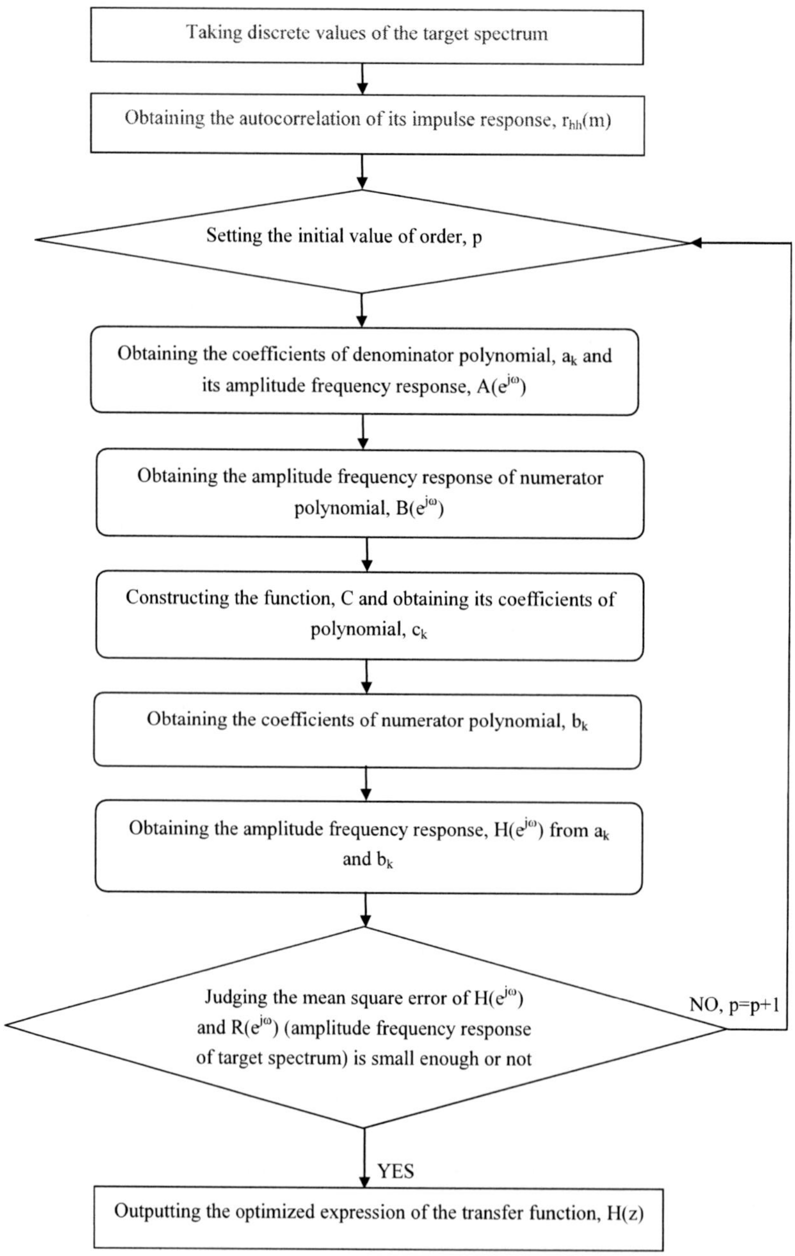

Figure 2. Flow chart for obtaining the optimized expression of transfer functions.

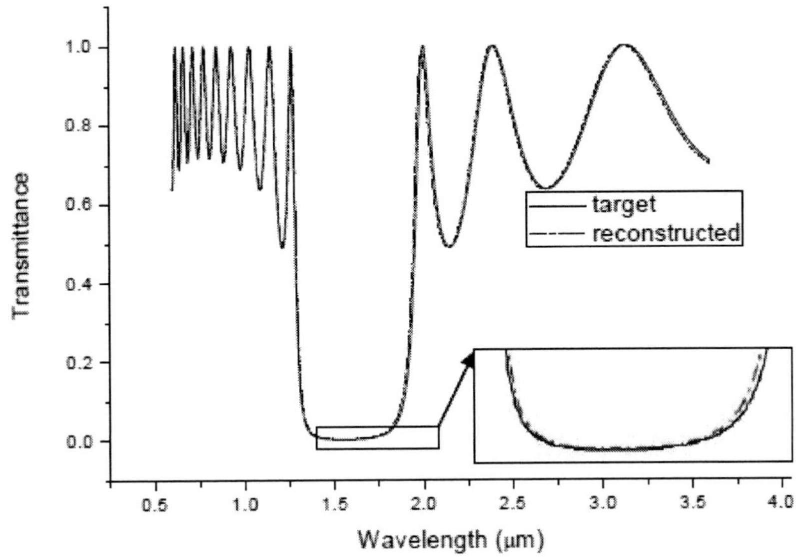

Figure 3. Transmittance spectrum of the $(HL)^5H$ 1DPC structure. The inset shows the enlarged view of the bottom part of the band gap.

Table 1. Reflection coefficients of each interface

Reflection coefficients	Value	Reflection coefficients	Value
r_{01}	- 0.4086	r_{67}	- 0.2570
r_{12}	0.2624	r_{78}	0.2592
r_{23}	- 0.2614	r_{89}	- 0.2556
r_{34}	0.2620	r_{910}	0.2576
r_{45}	- 0.2591	r_{1011}	- 0.2542
r_{56}	0.2608	r_{1112}	0.3928

Table 2. Refractive indices of each layer

Layer	Refractive index	Layer	Refractive index
0 (incident media)	1	7	2.3430
1	2.3819	8	1.3785
2	1.3915	9	2.3253
3	2.3764	10	1.3727
4	1.3898	11	2.3082
5	2.3619	12 (exit media)	1.0063
6	1.3848		

Compared with the values of the refractive index of each layer, the structural parameters obtained by the DSP method accord well with the original optical ones. The slight deviation originates from the numerical calculation and almost has no effect on the output stopband spectrum (the transmittance spectrum of the synthesized stopband structure is shown as the dash-dotted curve in Figure 3). The deviation can be further reduced by sampling the target spectrum with more discrete values, which increases the amount of computation in the same time. Its total amount of computation is similar to that of a forward simulation process (such as the transfer matrix method) and can often be fulfilled in few seconds with a personal computer.

This design method is also suited for the 1DPC with defect modes, which reflect the localized state characteristics of the photonic band gap. The target transmittance spectrum is shown as the solid curve in Figure 4. There is a narrow transmission peak at 1.55 μm in the forbidden band. The reflection transfer function in digital signal domain of the target spectrum can be calculated out as

$$H(z)=\frac{0.3919-0.6005z^{-1}+08892z^{-2}-1.1682z^{-3}+1.3935z^{-4}-1.5227z^{-5}+1.5211z^{-6}-1.3704z^{-7}}{1-1.0312z^{-1}+1.3244z^{-2}-1.5350z^{-3}+1.6213z^{-4}-1.5529z^{-5}+1.3217z^{-6}-0.9413z^{-7}}$$
$$\frac{+1.0711z^{-8}-0.6461z^{-9}-0.0007z^{-10}+0.6466z^{-11}-1.0705z^{-12}+1.3676z^{-13}-1.5166z^{-14}}{+0.4486z^{-8}+0.1026z^{-9}-0.6929z^{-10}+1.1486z^{-11}-1.3740z^{-12}+1.4594z^{-13}-1.4134z^{-14}}$$
$$\frac{+1.5163z^{-15}-1.3866z^{-16}+1.1610z^{-17}-0.8831z^{-18}+0.5956z^{-19}-0.3886z^{-20}}{+1.2608z^{-15}-1.0367z^{-16}+0.7802z^{-17}-0.5282z^{-18}+0.3097z^{-19}-0.1523z^{-20}}. \quad (11)$$

As shown in Equation (11) and according to the recursion algorithm, a 1DPC with 1/4λ $(HL)^5(LH)^5$ configuration is synthesized. The refractive indexes of H and L layers are about 2.3 and 1.4, which are nearly the values of TiO_2 (2.28) and SiO_2 (1.43) respectively. The transmittance spectrum of the synthesized structure almost coincides with the target one and is shown as the dash-dotted curve in Figure 4.

For the above design examples, the numbers of the sampling points are about 2000 and 20000 in the former and later target spectrum respectively. Computation can be fulfilled within 2-3 seconds with a personal computer (CPU @3.1GHz and 4 GB RAM). It should be pointed out that a large amount of computation will be needed for designing an arbitrarily given target band gap structure (no transmittance spectrum is obtained in advance). Firstly, a square wave-like transmittance spectrum (at least one period is included) containing the information about the location and the width of the band gap is set as the target.

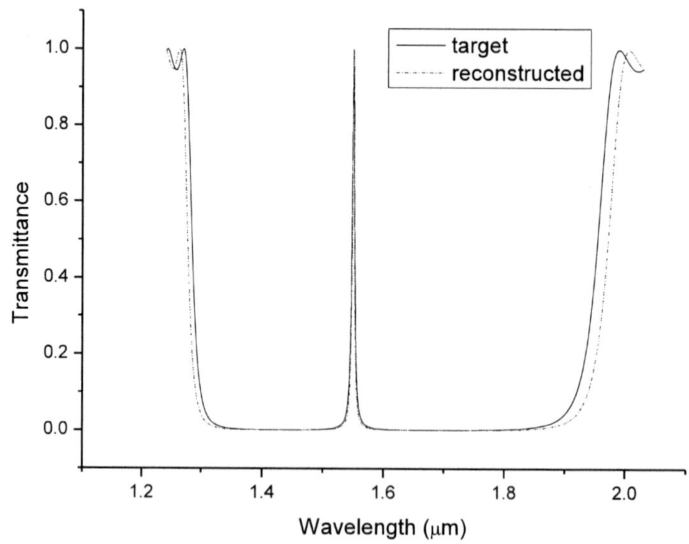

Figure 4. Transmittance spectrum of the $(HL)^5(LH)^5$ 1DPC structure.

And then the spectrum is sampled into discrete values. A set of reflection transfer function, reflection coefficients and the refractive index of each layer can be then searched out if a big deviation (such as a big ripple on both sides of the band gap as shown in Figure 3) is allowed. Afterwards, whether all the obtained refractive indexes have their own physical meaning should be determined. If any, a refractive index does not have the physical meaning (too small or too big, far from that of a real material), the numerical searching should be redone until a set of physically meaningful refractive indexes is obtained.

REFERENCES

[1] Joannopoulos J D, Villeneuve P R and Fan S, Photonic-bandgap microcavities in optical waveguides, *Nature*, vol. 386, pp. 143-145, 1997.

[2] Winn J N, Fink Y, Fan S and Joannopoulos J D, Omnidirectional reflection from a one-dimensional photonic crystal, *Optics Letters*, Vol, 23, pp. 1573-1575, 1998.

[3] Ozaki R, Matsui T, Ozaki M and Yoshino K, Electrically color-tunable defect mode lasing in one-dimensional photonic-band-gap system containing liquid crystal, *Applied Physics Letters*, Vol. 82, pp. 3593, 2003.

[4] Hopman W C L, Pottier P, Yudistira D, van Lith J, Lambeck P V, De La Rue R M, Driessen A, Hoekstra H J W M and de Ridder R M, Quasi-one-dimensional photonic crystal as a compact building-block for refractometric optical sensors, *IEEE Journal of Selected Topics in Quantum Electronics*, Vol. 11, pp. 11-16, 2005.

[5] Chen P Z, Hou G F, Suo S, Ni J, Zhang J J, Zhang X D and Zhao Y, One-dimensional photonic crystal(1D PC)-based back reflectors for amorphous silicon thin film solar cell, *Acta Phys. Sin.* Vol. 63, pp. 077301, 2014.

[6] Fan W L and Dong L F, Spontaneous transition of one-dimensional plasma photonic crystal's orientation in a dielectric barrier discharge, *Chin. Phys. B*, Vol. 22, pp. 0142 13, 2013.

[7] Fan C Z, Wang J Q, He J N, Ding P and Liang E J, Theoretical study on the photonic band gap in one-dimensional photonic crystals with graded multilayer structure, *Chin. Phys. B*, Vol. 22, pp. 074211, 2013.

[8] John G. Proakis and Dimitris G. Manolakis: Digital Signal Processing: pricinples, algorithm, and applications (New Jersey: Printice Hall), 1996.

[9] J. Zhang, X. Yang, Universal Michelson Gires-Tournois interferometer optical interleaver based on digital signal processing, *Optics Express*, vol. 18, pp. 5075-5088, 2010.

[10] J. Zhang, S. Guo, X. Li, General IIR optical notch filter based on Michelson Gires-Tournois interferometer, *Optics Communications*. vol. 285, pp. 491-496, 2012.

[11] Madsen C K and Zhao J H, *Optical Filter Design and Analysis: A Signal Processing Approach* (John Wiley and Sons), 1996.

[12] Cheng C H, Signal processing for optical communication, *IEEE Signal Processing Magazine*, Vol. 23, pp. 88-96, 2006.

[13] Skaar J, Wang L and T. Erdogan, Synthesis of thick optical thin-film filters with a layer-peeling inverse-scattering algorithm, *Applied Optics*, Vol, 40, pp. 2183, 2001.

[14] Skaar J and Waagaard O H, Design and characterization of finite-length fiber gratings, *IEEE Journal of Quantum Electronics* Vol. 39, pp. 1238-1245, 2003.

[15] Feced R, Zervas M N and Muriel M A, An efficient inverse scattering algorithm for the design of nonuniform fiber Bragg gratings, *IEEE J. Quantum Electron*, Vol. 35, pp. 1105-1115, 1999.

[16] Frolik J L and Yagle A E, An asymmetric discrete-time approach for the design and analysis of periodic waveguide gratings, *Journal of Lightwave Technology*, Vol. 13, 175-185, 1995.

[17] Bruckstein A M, Levy B C and Kailath T, Differential methods in inverse scattering, *SIAM J. Appl. Math,* Vol. 45, pp. 312-335, 1985.

[18] Yagle A E and Levy B C, Layer-stripping solutions of multidimensional inverse scattering problems, *J. Math. Phys*, Vol, 27, pp. 1701, 1986.

[19] Ingle V K and Proakis J G, *Digital Signal Processing Using MATLAB* (3rd edition, Cengage Learning), 2011.

[20] Graham P and Nelson B, Reconfigurable processors for high-performance, embedded digital signal processing, in Field Programmable Logic and Applications (Ed. P. Lysaght, J. Irvine, R. Hartenstein, Springer) pp. 1-10, 1999.

[21] Dowling E M and MacFarlane D L, Lightwave lattice filters for optically multiplexed communication systems, *Journal of Lightwave Technology,* Vol. 12, pp. 471-486, 1994.

In: Digital Signal Processing (DSP)
Editor: Juan Zhang

ISBN: 978-1-63485-168-8

Chapter 7

DIGITAL SIGNAL PROCESSING IN INTEFEROMETRIC OPTICAL INTERLEAVERS

Juan Zhang [*] ***and Xiaowei Yang***
School of Communication and Information Engineering,
Shanghai University, Shanghai, China

ABSTRACT

Optical interleavers, which can combine and/or separate comb signals, are important building blocks in the optical communication system. They can efficiently reduce the technical requirements of the dense wavelength-division multiplexed (DWDM) devices (such as filters) for smaller wavelength spacings. In this chapter, from the perspective of digital signal processing, optical interleavers based on Michelson Gires-Tournois interferometer structure and birefringent crystals structure are, respectively, proposed and analyzed in detail.

Keywords: optical interleaver, Michelson Gires-Tournois interferometer, birefringent crystal, digital signal processing

INTRODUCTION

To meet the explosive-growing demand for bandwidth, the dense wavelength division multiplexing (DWDM) systems should offer more channels with narrower channel spacings. However, this would increase the technical demands of all the filtering components used in the system. Optical interleaver, which can separate a high-channel density stream into two complimentary low-channel density streams (and also can be used in a reverse direction), is an important device to ease such a technical burden [1, 2]. Optical interleavers can be divided into two categories: symmetrical interleavers and asymmetrical interleavers [3]. Symmetrical interleavers separate (or combine) even and odd channels from the DWDM comb with two

[*] Corresponding author: Juan Zhang. Email: juanzhang@staff.shu.edu.cn.

symmetrical output spectra (the bandwidthes of even and odd channels are equal, duty cycle = 1:1). Interleavers with asymmetrical output spectra are called asymmetrical interleavers. In fact, a symmetrical interleaver can be considered as a special case of asymmetrical interleavers. Asymmetrical interleavers with arbitrary duty cycles can significantly enhance the flexibility and applicability of optical networks [4]. Main technical requirements of an interleaver include wide and flat passband, low dispersion, low insertion loss, high channel isolation and low thermal drift.

Up to now, a lot of technical schemes have been proposed to design optical interleavers. Most of them are symmetrical devices [5-14]. A few interferometric structues, such as Michelson-Gires-Tournois interferometer (MGTI) [3, 15], cascaded Mach–Zehnder interferometers (MZI) [16], birefringent MGTI [17], double stage symmetrical interleaver with wavelength shifting capability [18], and birefringent interleaver [19], etc. have been proposed to achieve asymmetric interleaving. Optical interleavers based on Michelson Gires-Tournois interferometer structure and birefringent crystals structure are focused on in this chapter from the perspective of digital signal processing.

Universal Michelson Gires-Tournois Interferometer Optical Interleavers

Michelson Gires-Tournois interferometer (MGTI) with cascaded reflectors can be designed to achieve symmetrical or asymmetrical interleaving. In this section, based on digital signal processing theory, design examples of symmetrical and asymmetrical interleavers with arbitrary duty cycles are given. The results are analyzed and compared in detail.

Figure 1 shows a schematic diagram of the MGTI interleaver. It is form by two sets of multi-cavity GTEs (MC-GTEs), a 50:50 beam splitter to output even channels, and an optical circulator to output odd channels.

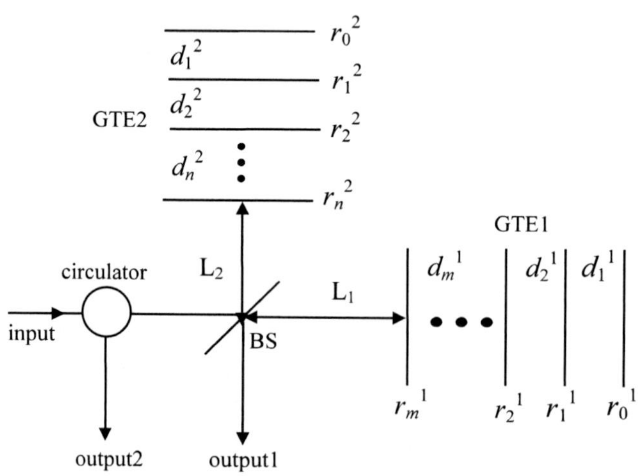

Figure 1. Schematic diagram of an MGTI interleaver.

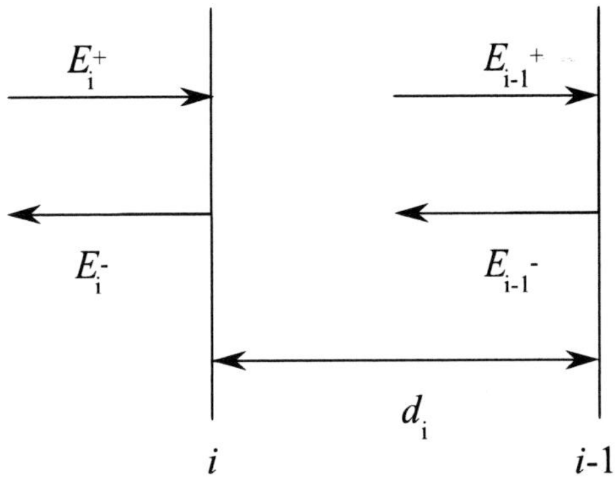

Figure 2. Amplitudes at reflectors *i* and *i*-1 out of a stack of multi-reflectors.

GTE1 and GTE2 consist of $m + 1$ and $n + 1$ reflectors respectively. Reflectance and reflection coefficients of the $m + 1$ reflectors are R_0^1, R_1^1, R_2^1, … , R_m^1, and r_0^1, r_1^1, r_2^1, … , r_m^1, respectively. Reflectance and reflection coefficients of the $n + 1$ reflectors are R_0^2, R_1^2, R_2^2, … , R_n^2, and r_0^2, r_1^2, r_2^2, … , r_n^2, respectively. Cavity length of GTE1 and GTE2 are d_1^1, d_2^1, … , d_m^1 and d_1^2, d_2^2, … , d_n^2, respectively. The reflectance R_0^1 and R_0^2 of the rear-reflector are assumed to be unity so that MC-GTE is a lossless all-pass filter in theory. L_1 and L_2 are the arm length of the MI.

Optical interleavers based on Michelson Gires-Tournois interferometer (MGTI) with arbitrary cascaded reflectors for symmetrical or asymmetrical periodic frequency response with arbitrary duty cycles are defined as universal MGTI optical interleaver (UMGTIOI). A novel and simple method based on digital signal processing can be utilized for the design of UMGTIOI.

Digital Filter Model

Figure 2 indicates amplitudes at reflectors i and i–1. E_i^+ and E_{i-1}^+ are, respectively, amplitude of the electric-field vector on the left side of reflectors i and i–1 for a wave front traveling to the right. E_i^- and E_{i-1}^- are, respectively, amplitude on the left side of reflectors i and i–1 for a wave front traveling to the left. The distance between reflectors i to i–1 is d_i.

From Figure 2, we obtain:

$$\begin{pmatrix} E_i^+ \\ E_i^- \end{pmatrix} = \frac{1}{t_i}\begin{bmatrix} e^{i\phi_i} & r_i e^{-i\phi_i} \\ r_i e^{i\phi_i} & e^{-i\phi_i} \end{bmatrix}\begin{pmatrix} E_{i-1}^+ \\ E_{i-1}^- \end{pmatrix}, \tag{1}$$

Therefore

$$\frac{E_i^-}{E_i^+} = \frac{r_i e^{j\phi_i} + e^{-j\phi_i}\left(E_{i-1}^- / E_{i-1}^+\right)}{e^{j\phi_i} + r_i e^{-j\phi_i}\left(E_{i-1}^- / E_{i-1}^+\right)} = e^{-j\Theta}, \tag{2}$$

where phase $\phi_i = 2\pi d_i/\lambda$, λ is optical wavelength, Θ is phase shift of GTE, and r_i and t_i are, respectively, amplitude reflection and transmission coefficients for reflector *i*. Phase shift of GTE with arbitrary reflectors can be obtained from Equation (2).

Amplitude transmission and reflection coefficients of the two output ports of the interleaver are

$$H_1 = \frac{1}{2}\{\exp[-j(\Theta_1 + 4\pi L_1/\lambda)] + \exp[-j(\Theta_2 + 4\pi L_2/\lambda)]\}, \tag{3}$$

$$H_2 = \frac{1}{2}\{\exp[-j(\Theta_1 + 4\pi L_1/\lambda)] - \exp[-j(\Theta_2 + 4\pi L_2/\lambda)]\}, \tag{4}$$

where Θ_1 and Θ_2 are phase shift of GTE1 and GTE2 respectively. Equations (3) and (4) can be re-written into

$$H_1 = \frac{1}{2}\exp(-i\,4\pi L_2/\lambda)[\exp(-i\Theta_1)\cdot\exp(-i\,4\pi\Delta L/\lambda) + \exp(-i\Theta_2)], \tag{5}$$

$$H_2 = \frac{1}{2}\exp(-i\,4\pi L_2/\lambda)[\exp(-i\Theta_1)\cdot\exp(-i\,4\pi\Delta L/\lambda) - \exp(-i\Theta_2)], \tag{6}$$

where $\Delta L = L_1 - L_2$ is difference between the two arm lengths. The term $\exp(-i\,4\pi L_2/\lambda)$ can be omitted because it does not affect intensity of the two output ports. Thus the MGTI interleaver is equivalent to a digital filter as shown in Figure 3. Here, $Z = \exp(i4\pi d/\lambda)$. In other words, $z^{-\Delta L/d}$ is an all-pass filter. $A_1(Z)$ and $A_2(Z)$ are the all-pass filter function corresponding to GTE1 and GTE2 respectively. *d* can be defined as reference cavity length and can be determined by $d = C/(2\cdot\Delta f)$, where *C* is the speed of light in vacuum and Δf is the period of optical spectrum.

Thus the transfer functions of the two output ports can be written as

$$H_1(z) = \frac{1}{2}[A_1(z)\cdot z^{-\frac{\Delta L}{d}} + A_2(z)], \tag{7}$$

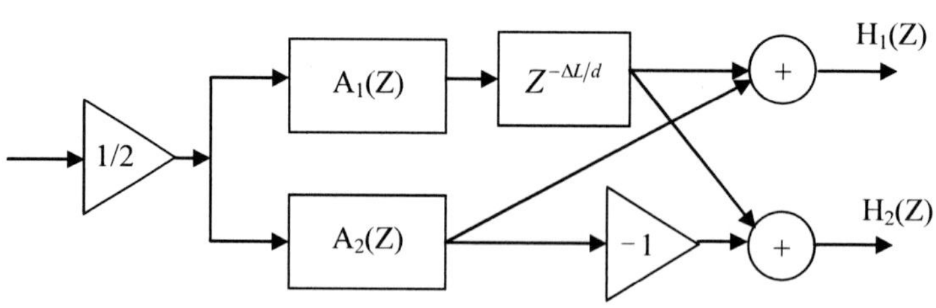

Figure 3. Digital filter model.

$$H_2(z) = \frac{1}{2}[A_1(z) \cdot z^{-\frac{\Delta L}{d}} - A_2(z)]. \tag{8}$$

Principle and Method of MGTI Interleaver Design

Because $A_1(z)$ and $A_2(z)$ are functions of z with real coefficients, the coefficients of $H_1(z)$ and $H_2(z)$ must be real. From the view point of digital signal processing, imaginary poles/ zeros of $H_1(z)$ and $H_2(z)$ are in conjugate pairs. Without loss of generality, a simplest case with one pair of conjugate imaginary pole/zero and one real pole/zero are analyzed. Assume that poles of $H_1(z)$ and $H_2(z)$ are all on the imaginary axis. Figure 4 shows a pole/zero plot. The zero z_0 is on the real axis, and zeros z_1 and z_1^* are a conjugate pair. Pole p_0 is located in the base point, and poles p_1 and p_1^* are a conjugate pair on the imaginary axis. Suppose $z = e^{j\omega}$, so

$$|H_1(e^{j\omega})| = \frac{|e^{j\omega} - z_1||e^{j\omega} - z_1^*||e^{j\omega} - z_0|}{|e^{j\omega} - p_1||e^{j\omega} - p_1^*||e^{j\omega} - p_0|}. \tag{9}$$

From Equation (9), it is clear that $|H_1(e^{-j\omega})| = |H_1(e^{j\omega})|$, meaning that $|H_1(e^{j\omega})|$ is symmetrical about ω = 0 and ω = π. Assume $d_1^1 = p_{11} \cdot d$, $d_2^1 = p_{12} \cdot d$, …, $d_m^1 = p_{1m} \cdot d$, $d_1^2 = p_{21} \cdot d$, $d_2^2 = p_{22} \cdot d$, …, $d_n^2 = p_{2n} \cdot d$, and $\Delta L = q \cdot d$, where $p_{11}, p_{12}, \ldots, p_{1m}, p_{21}, p_{22}, \ldots, p_{2n}$ and q are integers. The parameters $p_{11}, p_{12}, \ldots, p_{1m}, p_{21}, p_{22}, \ldots, p_{2n}$ and q are the ratio of the cavity length of the GTE1 ($d_1^1, d_2^1, \ldots, d_m^1$), GTE2 ($d_1^2, d_2^2, \ldots, d_n^2$) and the length difference of the two arms (ΔL) to the reference cavity length (d), respectively. Because $A_1(z)$ and $A_2(z)$ are all-pass filter functions, polynomials in the numerator and the denominator of $A_1(z)$ and $A_2(z)$ have the same coefficients, with the sequence in a reversed order. When $p_{11}, p_{12}, \ldots, p_{1m}, p_{21}, p_{22}, \ldots, p_{2n}$ are even and q is odd, the following relation is obtained:

$$\left|H_1(e^{j\frac{\pi}{2}})\right| = \left|H_1(e^{-j\frac{\pi}{2}})\right| = |\pm i \pm 1|/2 = \sqrt{2}/2. \tag{10}$$

It can be seen from Equation (10) that ω = π/2 and ω = −π/2 are 3dB frequencies. Therefore the 3dB bandwidth is π that just equals half of the period. From the above discussion, it can be concluded that poles of $H_1(z)$ (or $H_2(z)$) must be conjugate pairs on the imaginary axis to form a symmetric interleaver, otherwise an asymmetrical spectrum can be obtained.

As design of analog filter has been fully developed, one can design a digital filter by first designing an analog filter and then converting to a digital filter. Elliptical filters have equal ripple behavior in both the passband and stopband, and no other filter of an equal order can have a faster transition in gain between the passband and stopband [20]. Therefore the design method of elliptical filters is adopted here for the design of interleavers.

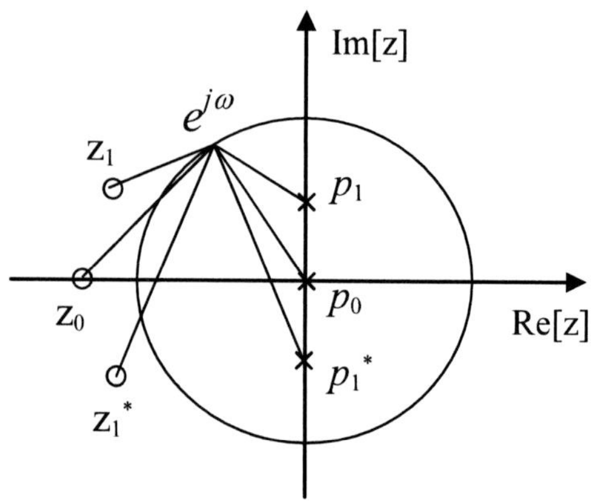

Figure 4. Schematic diagram of zero and pole locations.

Taking into account the above facts, we propose the following method to design interleavers with different bandwidth requirements. First, calculate $H_1(z)$ and $H_2(z)$ using Equation (7) and (8). Second, choose values of $p_{11}, p_{12}, \ldots, p_{1m}, p_{21}, p_{22}, \ldots, p_{2n}$ according to the requirements that the poles of $H_1(z)$ (or $H_2(z)$) are conjugate pairs on the imaginary axis for a symmetrical interleaver, and otherwise for an asymmetrical interleaver. Choose values of q to make $H_1(z)$ and $H_2(z)$ satisfy the format of the elliptic filter's transfer function. Thus the order of the interleaver, N, is determined. Third, design an N-th order elliptic filter which satisfies the design index of the interleaver. Finally, reflectance of all reflectors can directly be obtained by making the poles of $H_1(z)$ or $H_2(z)$ equal that of the N-th order elliptic filter. It should be pointed out that in order to make $r_0^1, r_1^1, r_2^1, \ldots, r_m^1$ and $r_0^2, r_1^2, r_2^2, \ldots, r_n^2$ to be real values, the $p_{11}, p_{12}, \ldots, p_{1m}, p_{21}, p_{22}, \ldots, p_{2n}$ and q should be integer. If $p_{11}, p_{12}, \ldots, p_{1m}, p_{21}, p_{22}, \ldots, p_{2n}$ and q are not integer, the above mentioned conditions can not be satisfied simultaneously, and the optimum design of the interleaver can not be accomplished.

Symmetrical Interleavers

As an example, GTE1 and GTE2 consisting of two reflectors are used to obtain a 50 GHz symmetrical interleaver with isolation ≤–30dB. Assume that $e^{-j4\pi \cdot d/\lambda} = e^{-j\omega} = z^{-1}$. R_0^1 and R_0^2 equal 1. The transfer functions of output ports can be obtained from Equations (2), (7) and (8) as follows:

$$H_1(z) = \frac{1}{2}\left[\frac{r_1^1 + z^{-p_{11}}}{1 + r_1^1 z^{-p_{11}}} \cdot z^{-q} + \frac{r_1^2 + z^{-p_{21}}}{1 + r_1^2 z^{-p_{21}}}\right], \tag{11}$$

$$H_2(z) = \frac{1}{2}\left[\frac{r_1^1 + z^{-p_{11}}}{1 + r_1^1 z^{-p_{11}}} \cdot z^{-q} - \frac{r_1^2 + z^{-p_{21}}}{1 + r_1^2 z^{-p_{21}}}\right]. \tag{12}$$

To assure that the poles of Equations (11) (and (12)) are conjugate on the imaginary axis, p_{11} and p_{21} can equal to 2. When q = 1, the transfer function of the output port satisfies the format of half-band elliptical filter. Thus the order of the interleaver is five. A fifth-order elliptic filter with isolation of –30dB is designed, with a pole/zero plot shown in Figure 5. Reflectance R_1^1 and R_1^2 is then obtained by making the poles of $H_1(z)$ or $H_2(z)$ equal that of the fifth-order elliptic filter. Its output spectrum is shown as the dot line in Figure 6. The solid and dash lines are output spectra obtained using the same method with isolation of –38 dB and –47 dB respectively. Design parameters corresponding to the three different output spectra are shown in Table 1. Figure 7 is an enlarged drawing in the passband of Figure 6. From Figures 6 and 7, it can be seen that isolation and passband width at 1 dB increase and decrease with the decrease of R_1^1 and R_1^2 respectively. Furthermore, a change on R_1^1 and R_1^2 has greater influence on isolation than on passband width. In practical applications, reflectance R_1^1 and R_1^2 should be determined according to practical demand.

Interleavers with other GTE structures can also be designed using the method in a similar manner as shown in Figure 8. Structures of the schemes in Figure 8 are (A) holophote GTE1 and GTE2, (B) two reflectors GTE1 and holophote GTE2, (C) two reflectors GTE1 and GTE2, (D) three reflectors GTE1 and two reflectors GTE2, and (E) three reflectors GTE1 and GTE2. Figure 8 (a) shows the output spectral transmittance of each scheme. Figures 8 (b) and 8 (c) are enlarged drawing in the passband and stopband respectively in one period.

Asymmetrical Interleavers

Take the structure of three reflectors GTE1 and two reflectors GTE2 as an example to design an interleaver with duty cycle 1:3. R_0^1 and R_0^2 equal 1.

Table 1. Design parameters for the 3 different spectra corresponding to the lines in Figures 6 and 7

Scheme	a	b	c
R_1^1	0.6022	0.4943	0.4147
R_1^2	0.0896	0.0512	0.0318

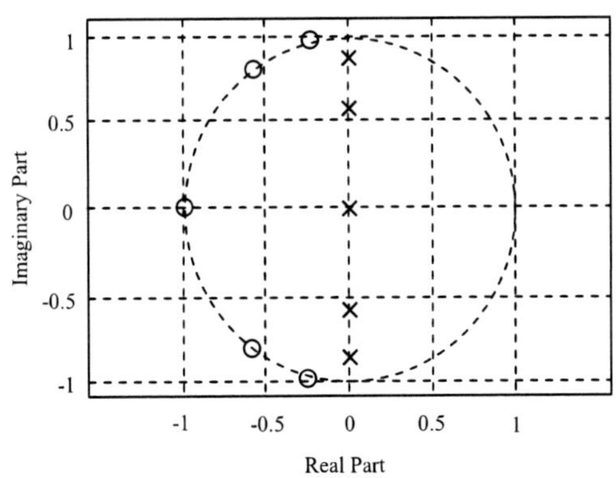

Figure 5. Pole/zero plot.

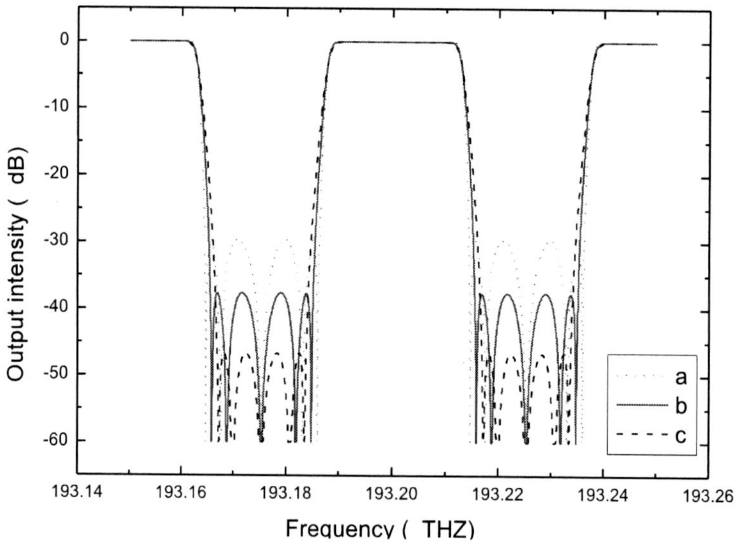

Figure 6. Output spectrum.

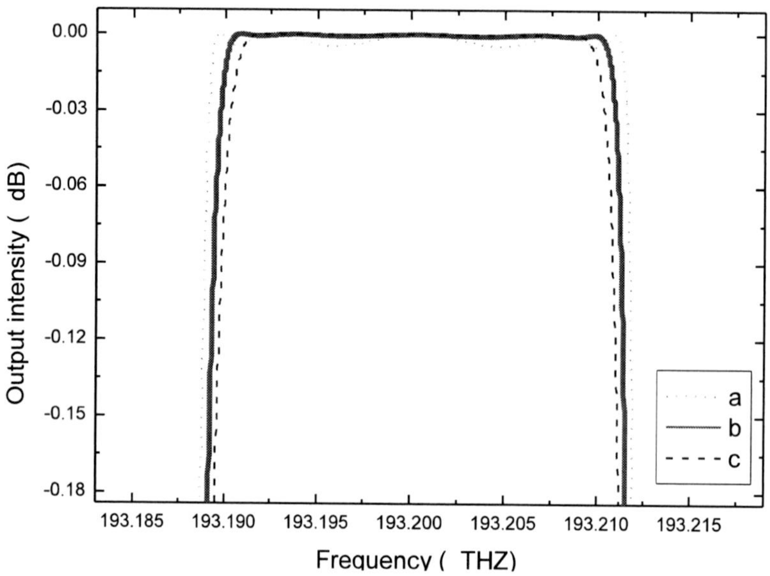

Figure 7. Zoomed spectra in the pass-band of Figure 6.

The transfer function of output ports can be obtained from Equations (2), (7) and (8) as follows:

$$H_1(z)=\frac{1}{2}\left[\frac{-r_2^{\,1}-r_1^{\,1}z^{-p_{12}}+r_2^{\,1}r_1^{\,1}z^{-p_{11}}+z^{-(p_{11}+p_{12})}}{1+r_2^{\,1}r_1^{\,1}z^{-p_{12}}-r_1^{\,1}z^{-p_{11}}-r_2^{\,1}z^{-(p_{11}+p_{12})}}\cdot z^{-q}+\frac{-r_1^{\,2}+z^{-p_{21}}}{1-r_1^{\,2}z^{-p_{21}}}\right], \tag{13}$$

$$H_2(z)=\frac{1}{2}\left[\frac{-r_2^{\,1}-r_1^{\,1}z^{-p_{12}}+r_2^{\,1}r_1^{\,1}z^{-p_{11}}+z^{-(p_{11}+p_{12})}}{1+r_2^{\,1}r_1^{\,1}z^{-p_{12}}-r_1^{\,1}z^{-p_{11}}-r_2^{\,1}z^{-(p_{11}+p_{12})}}\cdot z^{-q}-\frac{-r_1^{\,2}+z^{-p_{21}}}{1-r_1^{\,2}z^{-p_{21}}}\right]. \tag{14}$$

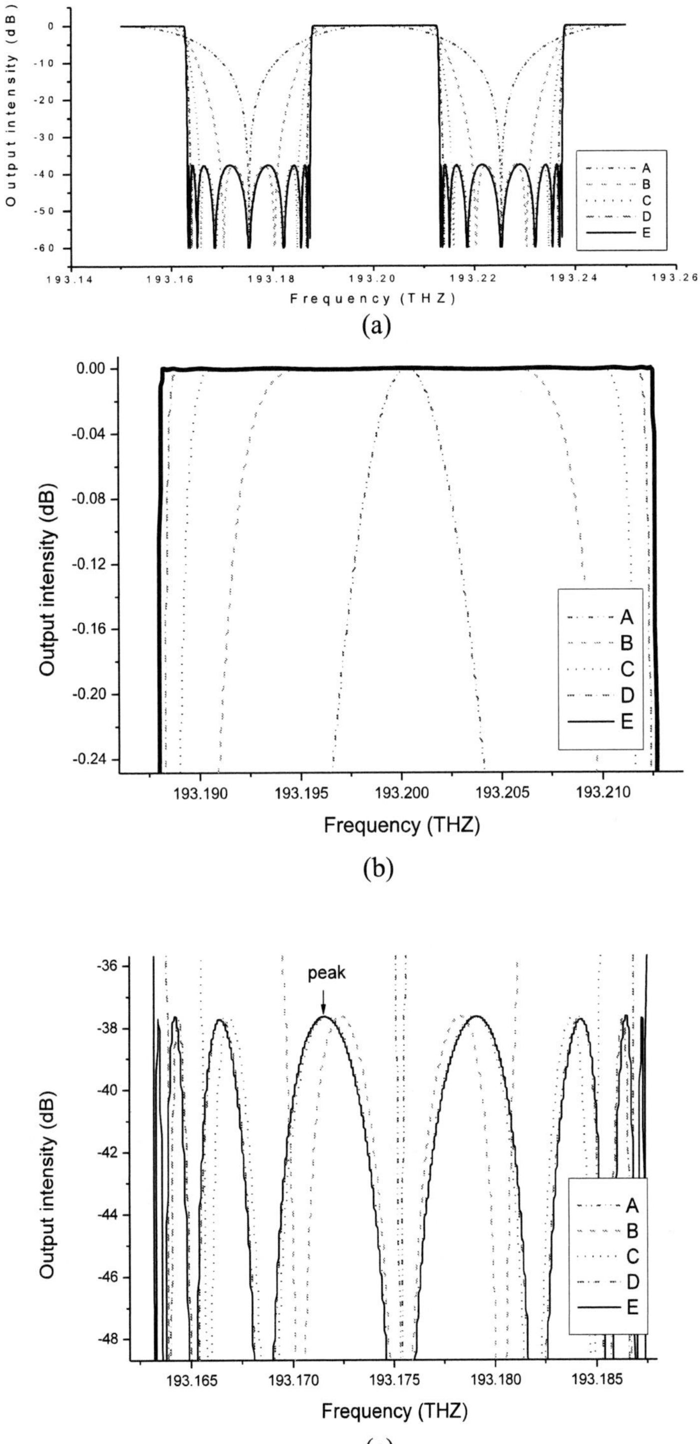

Figure 8. Output intensity spectra of the five schemes in different orders. (a) Output intensity spectra of five schemes. (b) Zoomed intensity spectra of pass-band. (c) Zoomed intensity spectra of stop-band.

From the design principle, p_{11}, p_{21} and p_{12} are equal to 1 and q is zero. Therefore, a third order elliptical filter should be designed. Such a filter that satisfies the requirement of spectral characteristics is designed, and its pole/zero plot shown in Figure 9. Making the poles of $H_1(z)$ or $H_2(z)$ equal to that of the third-order elliptic filter, the reflectance of reflectors can be directly obtained. Figure 10 shows intensity of the two output ports. The parameters are $R_1^1 = 0.2037$, $R_2^1 = 0.2468$, and $R_1^2 = 0.0709$.

From Figure 10, the two output spectra have isolation more than 35 dB. The passband width of wide and narrow port at 1dB is greater than 31GHz and 14GHZ respectively. An asymmetrical interleaver with other GTE structure can also be designed with the method. Figure 11 is the two output spectra of scheme A' (three reflectors GTE1 and holophote GTE2), C' (three reflectors GTE1 and two reflectors GTE2), and D' (three reflectors GTE1 and three reflectors GTE2). The output spectrum of [5] with two reflectors GTE1 and GTE2 structure (labeled B') is also drawn in Figure 11. In order to evaluate the spectral transmittance, passband bandwidth ratio is defined in [21]. It is the ratio of passband width at certain isolation to that at −3dB. According to the definition, passband bandwidth ratio for the four different schemes is shown in Table 2.

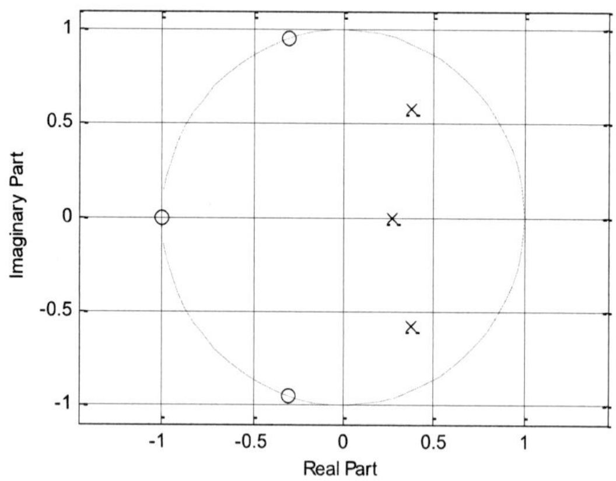

Figure 9. pole/zero plot.

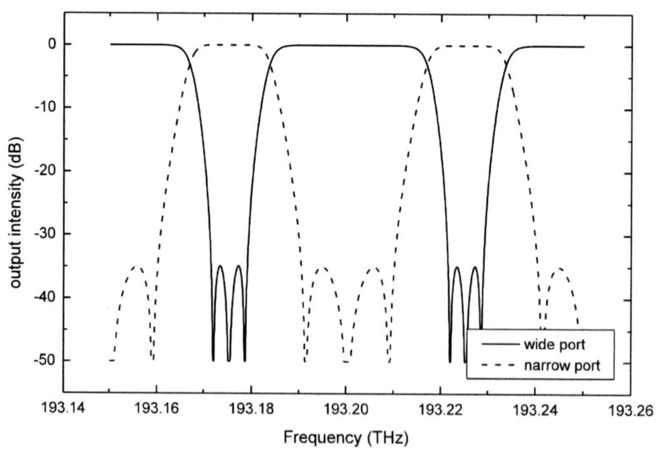

Figure 10. Output intensity spectrum of asymmetrical interleavers.

Table 2. Passband bandwidth ratio for different configurations

Schemes	Passband bandwidth ratio at −0.0436 dB isolation		Passband bandwidth ratio at −0.0173 dB isolation		Passband bandwidth ratio at −0.0069 dB isolation	
	Wide port	Narrow port	Wide port	Narrow port	Wide port	Narrow port
A'	66.2%	18.1%	61.7%	12.1%	58.1%	7.2%
B'	75.5%	26.5%	72.5%	18.1%	70.1%	11.5%
C'	78.7%	58.4%	74.9%	54.2%	71.9%	51.2%
D'	89.8%	68.1%	88.0%	64.5%	86.5%	61.5%

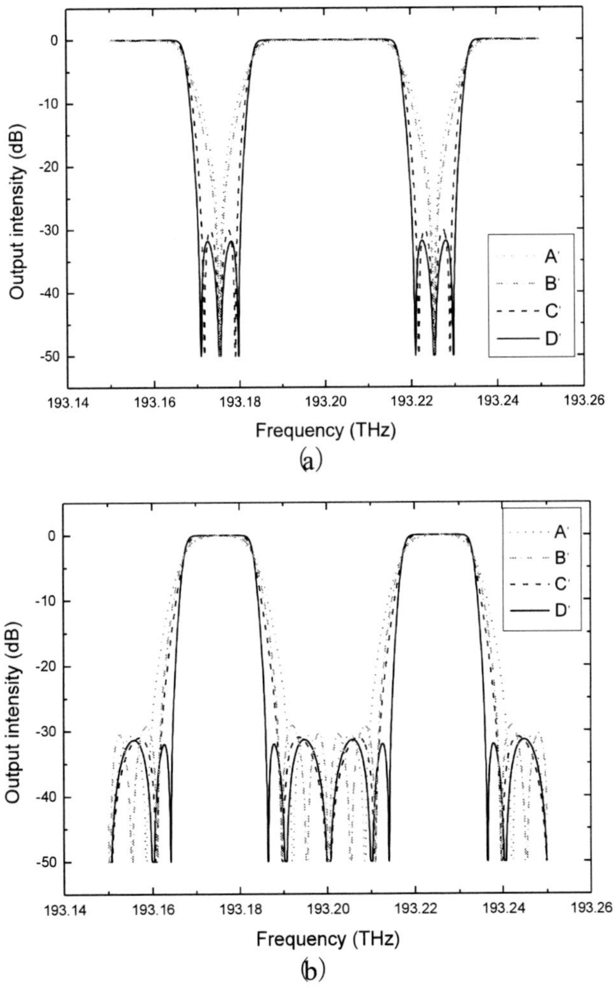

Figure 11. Output intensity spectra for two output ports obtained with different schemes. (a) wide port, (b) narrow port.

From Table 2 and Figure 11, passband width ratio increases with the increase of the reflector number. Scheme A' has the lowest passband band width ratio, whereas scheme D' has the highest. The passband bandwidth ratio of schemes B' and C' for wide port are similar, whereas that of the narrow port for schemes B' and C' is greatly different. That means the passband bandwidth ratio of narrow port can be greatly improved by replacing the two reflectors GTE with three reflectors GTE. For both wide port and narrow port, the change of passband bandwidth ratio at three different isolations is the largest for scheme A' and the smallest for scheme D'. Also, the rectangle degree of scheme A' is the lowest, and that of scheme D' is the highest.

Interleavers with Arbitrary Duty Cycles

An optical interleaver can be viewed as a lowpass-highpass doubly-complementary filter pair. In digital signal processing, an odd order lowpass-highpass doubly-complementary filter pair can always be decomposed into sum or difference of two all-pass filter with adjacent orders [22]. For example, a first-order filter can be decomposed into sum or difference of all-pass filter with order of 1 and 0, and a third-order filter can be decomposed into sum or difference of all-pass filter with order of 2 and 1. Therefore, from Equations (7) and (8), interleavers with arbitrary duty cycle can be accomplished by decomposing transfer functions into sum or difference of all-pass filter with adjacent orders when q equals zero. Parameters m and n related to the number of reflectors in GTE1 and GTE2 are determined by the orders of the two decomposed all-pass filter. Finally, reflectance of reflectors in the two GTEs can be obtained using the method described above.

In the previous subsection, a design example of third-order interleaver with duty cycle of 1:3 has been given with a structure of three reflectors GTE1 and two reflectors GTE2. From the principle, it can be designed to realize an interleaver with arbitrary duty cycles. Figure 12 shows the output spectrum of the interleaver with duty cycles 1:4, 1:5 and 1:6. The result of duty cycle 1:3 is also given for comparison. The design parameters are shown in Table 3.

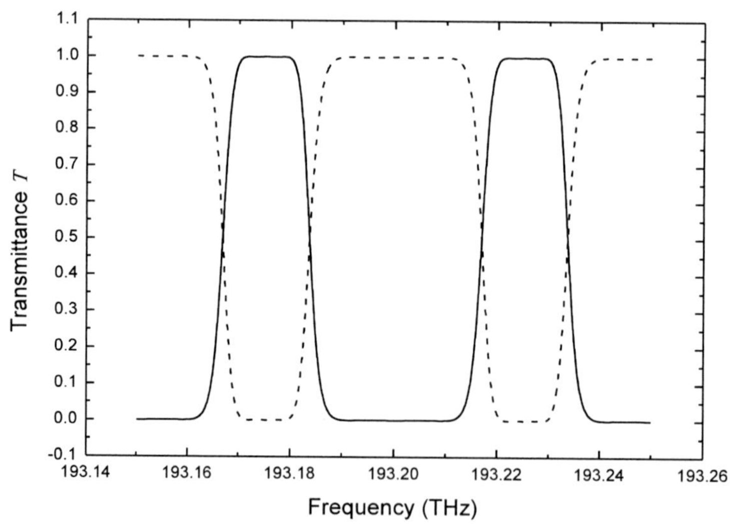

(a)

Figure 12. (Continued)

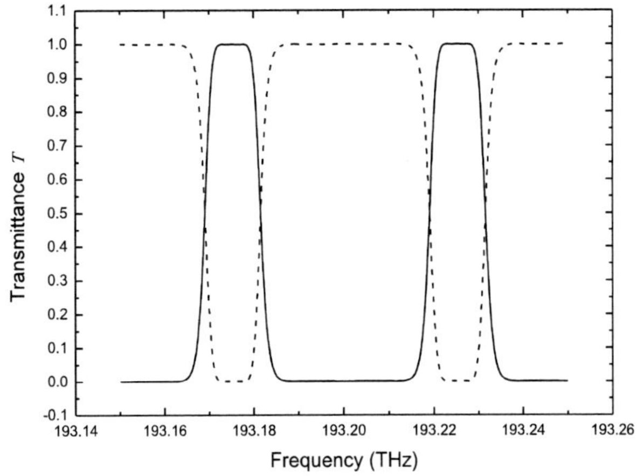

(b)

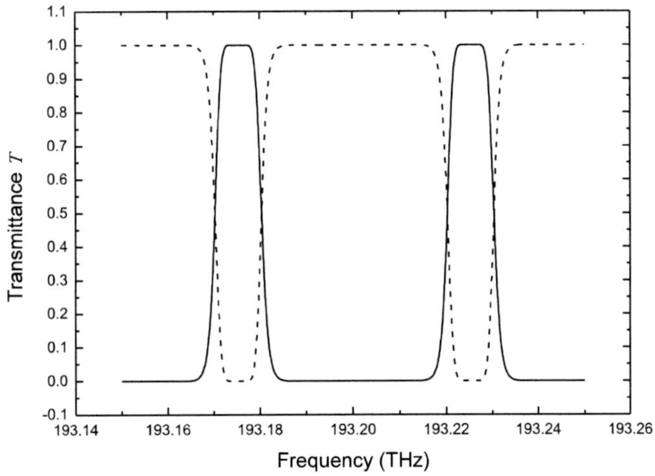

(c)

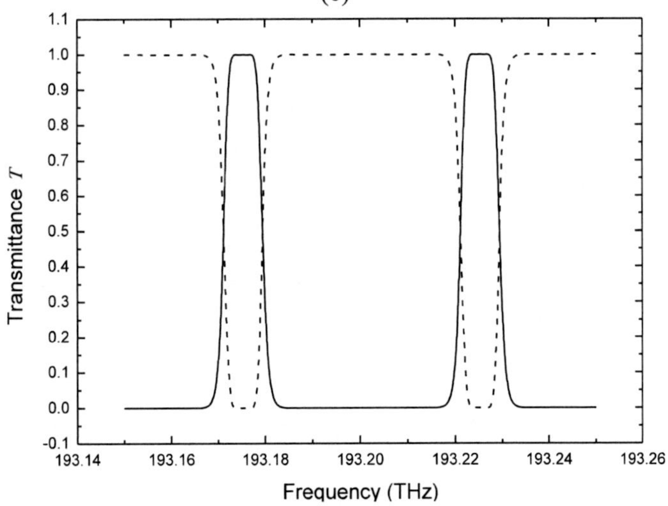

(d)

Figure 12. Output intensity spectra of four different duty cycles obtained with same structure. (a) duty cycle 1:3, (b) duty cycle 1:4, (c) duty cycle 1:5, (d) duty cycle 1:6.

Table 3. Design parameters for different duty cycles

Duty cycle	1:4	1:5	1:6
Design parameters	$R_1^1 = 0.2802$ $R_2^1 = 0.5010$ $R_1^2 = 0.1721$	$R_1^1 = 0.3476$ $R_2^1 = 0.6488$ $R_1^2 = 0.2559$	$R_1^1 = 0.4142$ $R_2^1 = 0.7520$ $R_1^2 = 0.3354$

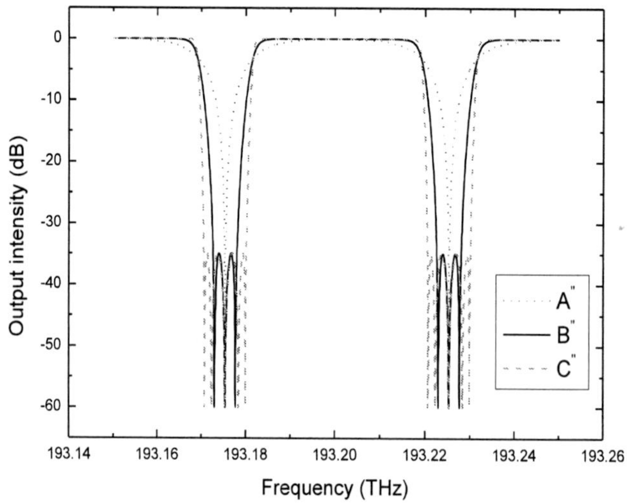

(a)

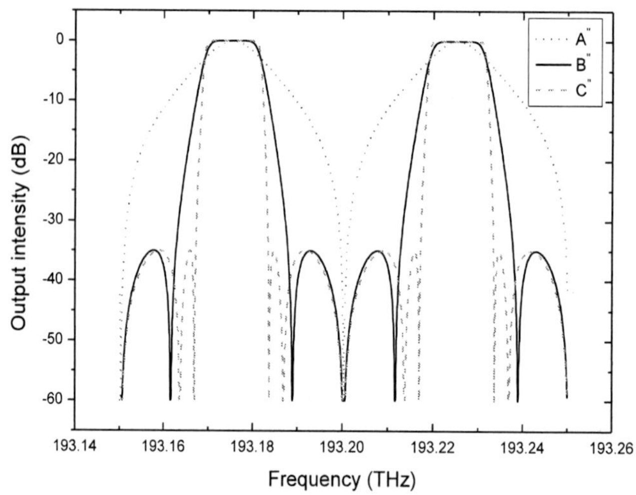

(b)

Figure 13. Output intensity spectra with duty cycle 1:4 obtained with three different structures. (a) wide port, (b) narrow port.

Interleavers of other orders with arbitrary duty cycle can also be designed in the same way. Figure 13 shows spectral transmittance of first-order and fifth-order interleavers with duty cycle 1:4. The design parameters are given in Table 4. The result of a third-order interleaver is also shown for comparison.

Table 4. Parameters of three different structures

Schemes	A": two mirrors GTE and holophote (first-order)	B": three mirrors GTE and two mirrors GTE (third-order)	C": Four mirrors GTE and three mirrors GTE (fifth-order)
Design parameters	R_1^1= 0.1683	R_2^1 = 0.2802, R_1^1 = 0.5010, R_1^2 = 0.1721	R_3^1 = 0.1094, R_2^1 = 0.7420, R_1^1 = 0.5663, R_2^2 = 0.1611, R_1^2 = 0.5003

From Figure 13, we can see that the rectangle degree of output spectrum can be improved by increasing the filter order. Output spectrum with higher order and better rectangle degree can be designed using the method. In the mean time, however, the fabrication complexity will increase. Therefore an appropriate structure should be chosen in practical applications.

OPTICAL INTERALEAVERS BASED ON BIREFRINGENT CRYSTAL STRUCTURE

Birefringent interleaver features benefits such as mature workmanship, low insertion loss, high channel isolation, and temperature compensation by different crystal wave plates. In this section, the mature digital signal processing technique finite impulse response (FIR) is adopted to design birefringent interleaver. A design result of 50-GHz asymmetrical interleaver based on three crystals is given. The passband bandwidth of the interleaver on wide port and on narrow port are respectively more than 60 GHZ, 30 GHZ, and the two output spectrums have wide flat-top passband and high channel isolation, it satisfies the transmission system of 40 Gb/s and 10 Gb/s.

Theory

The birefringent interleaver is composed of several cascaded crystal wave plates located between a polarizer and an analyzer. As shown in Figure 14. The polarization direction of polarizer is X axis. The angles between crystal optical axis and X axis are defined as crystals' azimuth angles, which are expressed as $\theta_1, \theta_2, \cdots\theta_i$ respectively. The angle between polarization direction of analyzer axis and X axis is defined as analyzer' azimuth angle, which is expressed as θ_p. The thickness of each crystal is L, Δn is the difference of crystal's refractive index. $k = 2\pi / \lambda$, λ is wavelength.

Assumed that there is no reflection of light from either surface of the crystal wave plate and the light is totally transmitted through the plate surface. According to Jones-matrix method, we can write the whole system as follows.

$$F = P_2 G_N \cdots G_1 P_1 \tag{15}$$

where P_1 and P_2 are, respectively, Jones-matrixes of polarizer and an analyzer, $G_1, G_2, \cdots, G_N$ are, respectively, the Jones-matrixes of the 1st, 2nd, $\cdots$, the *N*th crystal wave plate.

Define $\varphi_1 = \theta_1$, $\varphi_2 = \theta_2 - \theta_1$, $\varphi_3 = \theta_3 - \theta_2$, $\cdots, \varphi_{N+1} = \theta_p - \theta_N$; $c_i = \cos\varphi_i$, $s_i = \sin\varphi_i$, Equation (15) can be expressed as

$$F = \begin{bmatrix} c_{N+1} z^{-1/2} & s_{N+1} z^{1/2} \\ -s_{N+1} z^{-1/2} & c_{N+1} z^{1/2} \end{bmatrix} \begin{bmatrix} c_N z^{-1/2} & s_N z^{1/2} \\ -s_N z^{-1/2} & c_N z^{1/2} \end{bmatrix} \cdots \begin{bmatrix} c_2 z^{-1/2} & s_2 z^{1/2} \\ -s_2 z^{-1/2} & c_2 z^{1/2} \end{bmatrix} \begin{bmatrix} c_1 z^{-1/2} & s_1 z^{1/2} \\ -s_1 z^{-1/2} & c_1 z^{1/2} \end{bmatrix} \tag{16}$$

where $z^{-1} = e^{-j\omega} = e^{-jk\Delta nL}$. Assumed that the input signal of light is $\begin{bmatrix} 1 \\ 0 \end{bmatrix}$, the output signal of light passing through the interleaver filter can be expressed as

$$E_{out} = F \cdot \begin{bmatrix} 1 \\ 0 \end{bmatrix} = \begin{bmatrix} F1 \\ F2 \end{bmatrix}. \tag{17}$$

The power of *F*1 and *F*2 is complementary according to the conversation of energy. Equation (17) can be simplified as follows:

$$F1 \cdot z^{-N/2} = a_0 + a_1 z^{-1} + a_2 z^{-2} + \cdots + a_i z^{-i} + \cdots + a_N z^{-N}, \tag{18}$$

where $a_0, a_1, \cdots, a_N$ are the combined sum, difference and sum-and-difference terms of $c_1, c_2, \cdots, c_{N+1}$ and $s_1, s_2, \cdots, s_{N+1}$. We can see that the right of the Equation (18) satisfies the expression of FIR in digital signal processing. Considering that $z^{-N/2}$ doesn't influence magnitude-frequency characteristic, only changes phase-frequency characteristic, therefore, *F*1 can be obtained through approximating target filtering function by utilizing the FIR methods. Based on the obtained expressions of *F*1, *F*2 can then be obtaind, and finally the azimuth angles of each crystal and analyser can be obtained.

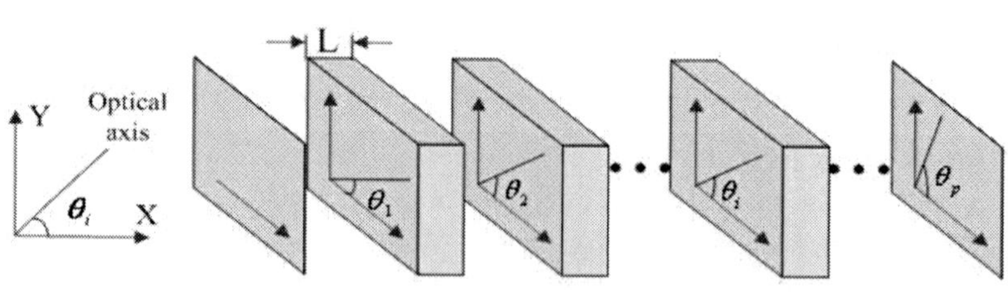

Figure 14. The structure of birefringent interleaver.

Results and Discussions

In digital signal processing, we can obtain the $F1$ by using multiple methods such as window, least-squares, equiripple, etc. Here equiripple method is used. This method can get better performance for both the passband and stopband. Seven-orders of $F1$ can be obtained by the equiripple method as followings:

$$F1 = -0.0341 - 0.0093\,z^{-1} + 0.1658z^{-2} + 0.3775z^{-3} + 0.3775z^{-4} + 0.1658z^{-5} - 0.0093z^{-6} - 0.0341z^{-7} . \quad (19)$$

Then $F2$ can be obtained according the complementary of the power of $F1$ and $F2$. Based the obtained expressions of $F1$ and $F2$, the azimuth angles of each crystal and analyser can be obtained with method of undetermined coefficients [23]. The results are shown in Table 5.

As azimuth angles of the first two, the middle three and the last two crystals are relatively close, they can be merged and modified. The analyser's azimuth angle remains the same. The modified azimuth angle of the first, the second and the third cystals are -63.0998°, -135.3269° and -207.4795° respectively. The thickness ratio of the three crystals equals to 2:3:3. The transmittance spectal of the birefringent interleaver with these parameters is shown in Figure 15. The 3dB bandwidth of the two output ports are 32.7 GHz and 67.3 GHz repectively. The isolation is more than 32dB.

The azimuth angles of the crystals and the analyser, and the thickness of the crystals are key factors determing the waveform, so it is important to discuss the influence of the change of the parameters on the waveform.

Figure 16 shows the spectral transmittance as a function of frequency when the azimuth angles θ_1, θ_2, θ_3 and θ_p are changed, respectively. It can be seen from Figure 16 that the change of θ_1, θ_2 and θ_3 is mainly exhibited in channel isolation and the passband bandwidth of narrow port. They have less impact on output spectrum of wide port. Whereas, the change of θ_p has more impact on channel isolation than on passband bandwidth of both narrow and wide port.

Figure 17 shows the spectral transmittance as a function of frequency when the thickness of the first, the second and the third crystal is changed, respectively. Here the change of thickness of crystal is replaced by that of the frequency spacing. The change of frequency spacing for the three curves in each subgrap of Figure 17 equals 0.002 GHz. The curves with the change of frequency spacing of zero are corresponding to case of structure with ideal parameters.

Table 5. Azimuth angles of the crystals and the analyser

crystal1	crystal 2	crystal 3	crystal 4
-63.4276	-61.6923°	-134.7095°	-134.4564°
crystal 5	crystal 6	crystal 7	analyser
-134.2034°	-207.2206°	-205.4853°	-180.8999°

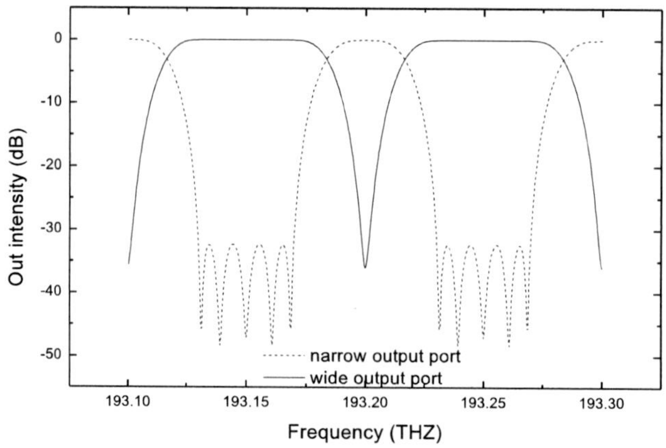

Figure 15. Output spectrum of the asymmetrical birefringent interleaver.

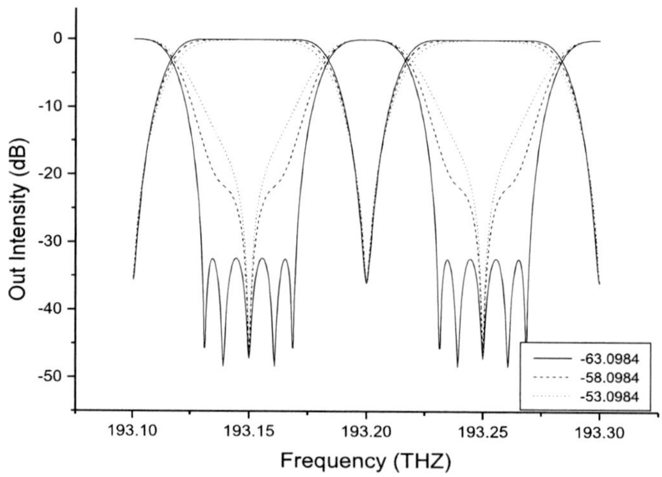

(a)

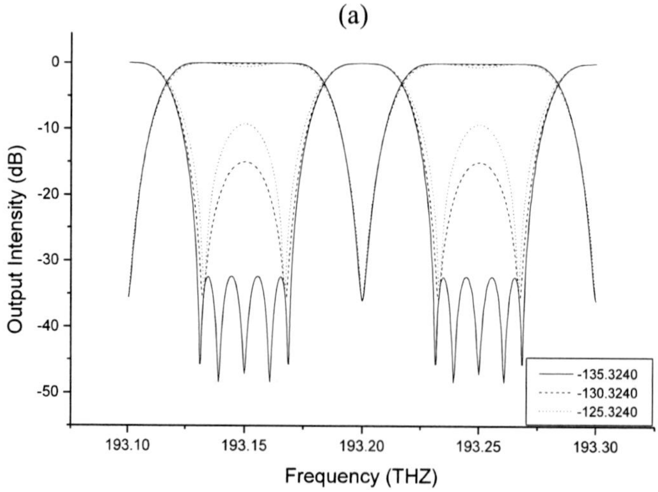

(b)

Figure 16. (Continued)

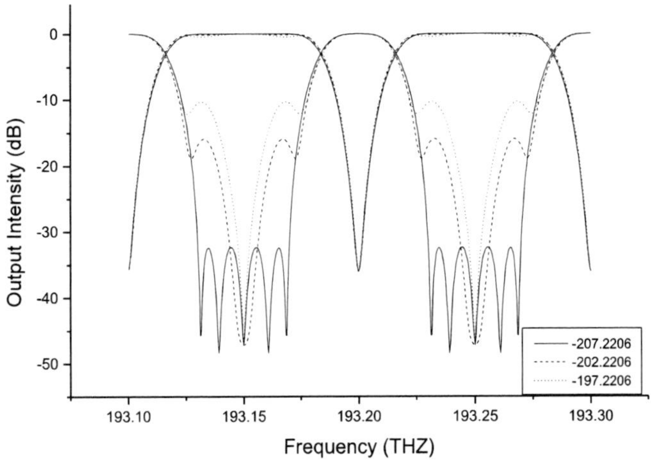

(c)

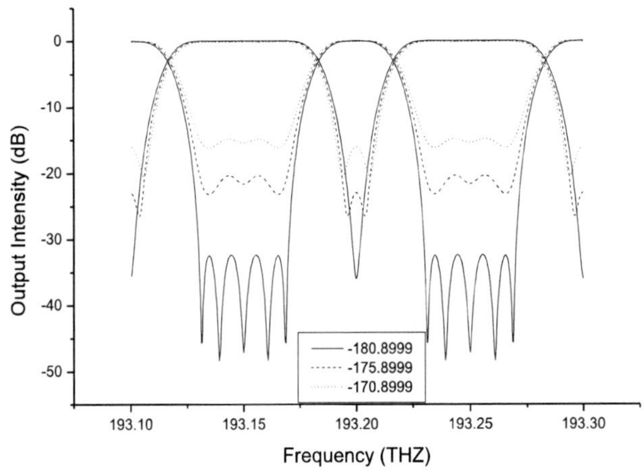

(d)

Figure 16. The spectral transmittance as a function of frequency for different azimuth angles of crystals. (a) When θ_2 = -135.3269°, θ_3 = -207.4795°, θ_p = -180.8999°, the waveform with different θ_1; (b) When θ_1 = -63.0998°,θ_3 = -207.4795° and θ_p = -180.8999°, the waveform with different θ_2; (c) When θ_1 = -63.0998°, θ_2 = -135.3269° and θ_p = -180.8999°, the waveform with different θ_3; (d) When θ_1 = -63.0998°, θ_2 = -135.3269° and θ_3 = -207.4795°, the waveform with different θ_p.

We can see from Figure 17 that the change of thiksness of the three crystals is mainly exhibited in channel isolation. It has less effect on the passband bandwidth of both the narrow and wide port.

Comparing Figure 16 with Figure 17, it can be found that the crystal thickness would bring larger influence on the spectral transmittance than that brought by the the azimuth angles. Thus the needed precision of the thickness is more rigorous than that of the azimuth angles. A little change of the thickness would bring larger influence on the spectral transmittance than that brought by the change of the azimuth angles.

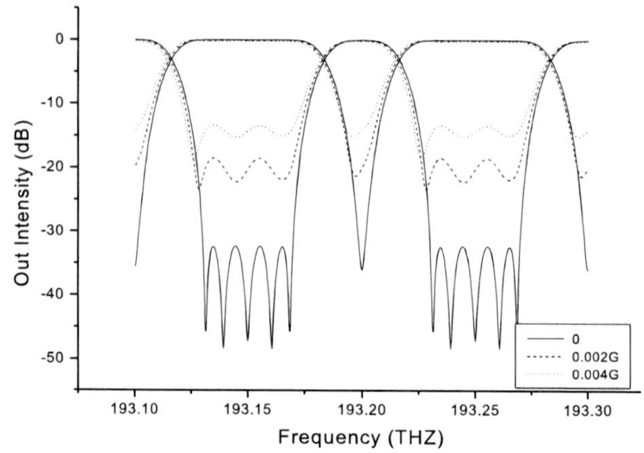

(a)

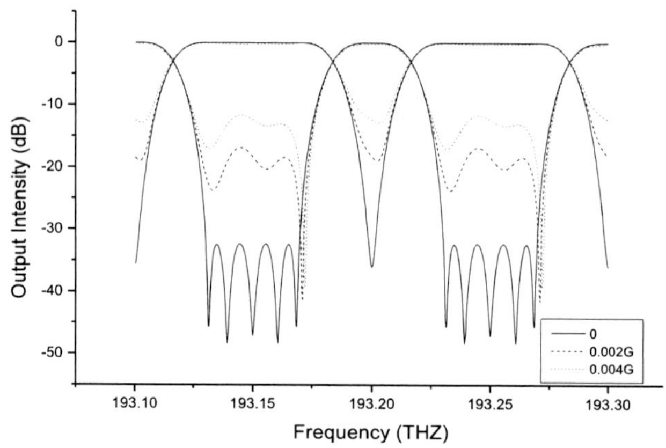

(b)

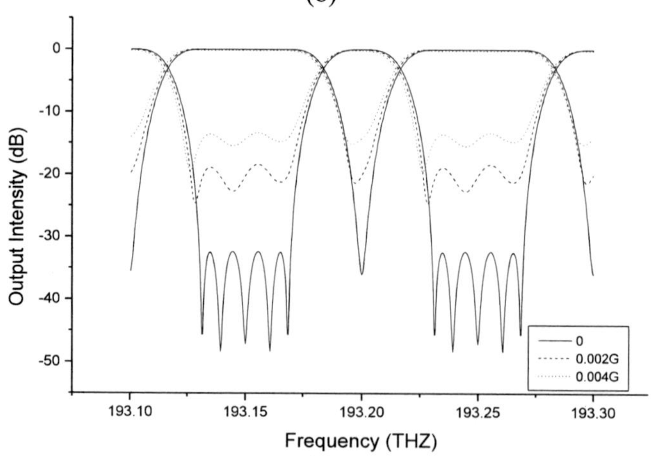

(c)

Figure 17. When θ_1 = -63.0998°, θ_2 = -135.3269°, θ_3 = -207.4795°, and θ_p = -180.8999°, the spectral transmittance as a function of frequency for (a) different frequency spacing of the first crystal; (b) different frequency spacing of the second crystal; (c) different frequency spacing of the third crystal.

REFERENCES

[1] B. Shine and J. Bautista, Interleavers make high-channel-count system economical, *Lightwave*, vol. 8, pp. 140-144, 2000.

[2] S. Cao, J. Chen, J. N. Damask, C. R. Doerr, L. Guiziou, G. Harvey, Y. Hibino, H. Li, S. Suzuki, K. Y. Wu, and P. Xie, Interleaver technology: comparisons and applications requirements, *IEEE/OSA J. Lightw. Technol.*, vol. 22, pp. 281-289, 2004.

[3] J. Zhang, X. Yang, Universal Michelson Gires-Tournois interferometer optical interleaver based on digital signal processing, *Optics Express*, vol. 18, pp. 5075-5088, 2010.

[4] B. Dingel, A. Dutta, Photonic add-drop multiplexing perspective for next generation optical networks (invited paper), *Proc. SPIE*, vol. 4532, pp. 394-408, 2001.

[5] J. Zhang, L. Liu, Novel Mach-Zehnder interferometer structure for tunable optical interleaver, *Optical Engineering*, vol. 45, pp. 045003, 2006.

[6] J. Zhang, Li. Liu, Y. Zhou, Optimum design of a novel electro-optically tunable birefringent interleaver filter, *Journal of Optics A: Pure and Applied Optics*, vol. 6, pp. 1052-1057, 2004.

[7] J. Zhang, Li. Liu, Y. Zhou, Novel and simple approach for designing lattice-form interleaver filter. *Optics Express*. vol. 11, pp. 2217-2224, 2003.

[8] J. Zhang, L. Liu, Y. Zhou, C. Zhou. Flattening spectral transmittance of birefringent interleaver filter. *Journal of Modern Optics*. vol. 50, pp. 2031-2041, 2003.

[9] J. Zhang, L. Liu, Y. Zhou, A tunable interleaver filter based on analog birefringent units. *Optics Communications*. vol. 227, pp. 283-294, 2003.

[10] J. Zhang, L. Liu, Y. Zhou, C. Zhou. Dynamic characteristics of a novel flat-top interleaver filter. *Optik*. vol. 114, pp. 39-44, 2003.

[11] C. H. Hsieh, C. W. Lee, S. Y. Huang, R. Wang, P. Yeh, and W. H. Cheng, Flat-top and low-dispersion interleavers using Gires–Tournoisetalons as phase dispersive mirrors in a Michelson interferometer, *Opt. Commun*, vol. 237, pp. 285-293, 2004.

[12] Z. P. Wang, S. J. Chang, C. Y. Ni, and Y. J. Chen, A high-performance ulrtacompact optical interleaver based on double-ring assisted Mach–Zehnder interferometer, *IEEE Photon. Technol. Lett.*, vol. 19, pp. 1072-1074, 2007.

[13] M. Gad, J. Ackert, D. Yevick, L. Chrostowski, and P. E. Jessop, Ring resonator wavelength division multiplexing interleaver, *J. Lightw. Technol.*, vol. 29, pp. 2102-2108, 2011.

[14] T. Mizuno, Y. Hida, T. Kitoh, M. Kohtoku, M. Oguma, Y. Inoue, and Y. Hibino, 12.5-GHz spacing compact and low-loss interleave filter using 1.5% silica-based waveguide, *IEEE Photon. Technol. Lett.*, vol. 16, pp. 2484-2486, 2004.

[15] C. H. Cheng, Asymmetrical interleaver structure based on the modified Michelson interferometer, *Opt. Eng*, vol. 44, pp. 115003, 2005.

[16] H.W. Lu, Y. Wei, B. G. Zhang, K. J. Wu, and G.W. Luo, Study of all fiber asymmetric interleaver based on Mach–Zehnder interferometer, *Opt. Commun.*, vol. 285, pp. 1118-1122, 2012.

[17] H. Hu, B. Zheng, Q. Liu, Y. Li, Li Wu, and S. Gu, Design of a birefringent Michelson interferometer-based interleaver with ultra-low dispersion and low cost, *Opt. Exp.*, vol. 19, pp. 24210-24218, 2011.

[18] N. J. Philip, A. Dogariu, L. Xu, L. Zong, T. Wang, O. Matsuda, and Y. J. Abe et al., Optical tunable asymmetric interleaver, In: *Proc. OFC*, 2006, Paper OTuM7.

[19] X. Yang, J. Zhang, Optimum design of asymmetric birefringent interleaver based on FIR digital filter design technique, *Proceedings of 2008 China-Japan Joint Microwave Conference (CJMW2008)*, pp. 595-598, 2008.

[20] J. G. Proakis, *Digital signal processing: Principles, Algorithms, and Applications*, (Prentice-Hall, New Jersey, 1996).

[21] Chao–Hsing Hsieh, Ruibo Wang, Zhiqing James Wen et al., Flat-top interleavers using two Gires-Tournois etalons as phase-dispersion mirrors in a Michelson interferometer, *IEEE Photon. Technol. Lett.*, vol. 15, pp. 242-244, 2003.

[22] S. K. Mitra, *Digital Signal Processing: a Computer-Based Approach*, (McGraw-Hill, New York, 2006).

[23] Kaname Jinguji, Masao Kawachi, Synthesis of coherent two-port lattice-form optical delay-line circuit, *Journal of Lightwave Technology*. vol. 13, pp. 73-82, 1995.

In: Digital Signal Processing (DSP)
Editor: Juan Zhang

ISBN: 978-1-63485-168-8

Chapter 8

Design of Infinite Impulse Response (IIR) Optical Notch Filter by Digital Signal Processing Method

Juan Zhang* and Xuguang Mao

[1]School of Communication and Information Engineering,
Shanghai University, Shanghai, China

[2]College of Electrical and Information Engineering,
Hunan University, Changsha, China

Abstract

Optical notch filters which can block undesired signals but do nothing to the useful signals are very useful. These filters are widely used in communication, biomedical engineering and other fields. In this chapter, optical notch filters based on Michelson Gires-Tournois interferometer structure and Mach-Zehnder interferometer (MZI) with double ring resonator coupled structure are proposed and analyzed respectively in detail from the perspective of digital signal processing.

Keywords: optical notch filter, optical allpass filter, Michelson interferometer, Gires-Tournois etalon, Mach-Zehnder interferometer, ring resonator, Z-transform

Introduction

Optical filters are important components of optical systems. Among them, notch filters which can block undesired signals but do nothing to the useful signals are very useful. These filters are widely used in communication, biomedical engineering and other fields [1-4].

Optical notch filter (ONF) can be realized by cascading optical ring resonators [5], birefringent crystals [6] and other structures [7], etc. Michelson Gires-Tournois interferometer

* Corresponding author: Juan Zhang. Email: juanzhang@staff.shu.edu.cn.

(MGTI) and Mach-Zehnder interferometer (MZI) are two classical interferometers of optical systems. They are characterized by simple structure, compact size, low cost, and high performance, therefore often used in the design of optical filters [8-11]. This chapter focuses on the Michelson Gires-Tournois interferometer structure and Mach-Zehnder interferometer (MZI) with double ring resonator coupled structure. From the perspective of digital signal processing, optical notch filters based on the two structures are, respectively, proposed and analyzed in detail.

General Optical Notch Filter Based on Michelson Gires-Tournois Interferometer

In this section, a general optical notch filter design is presented from a digital filter design perspective for Michelson Gires-Tournois Interferometer structure [12]. Optical notch filter with arbitrary notch frequency, notch point number, and 3 dB rejection bandwidth can be designed easily. Design examples of notch filters with one notch point, two notch points with equal and unequal 3 dB rejection bandwidth are presented. The change of output spectrum is also investigated for the reflectance of the reflectors and the distance between the reflectors deviating from the ideal value. The results show that the notch filter has the tunability of notch frequency and 3 dB rejection bandwidth. The chromatic dispersion characteristic of the notch filter is analyzed finally. It shows that the notch filter has excellent chromatic dispersion characteristic.

Design Principle

In frequency domain, the design of infinite impulse response (IIR) notch filter can be transformed into allpass filter [13]. The principle can also be used to design filters in optical domain. The basic theory is given in Figure 1.

From Figure 1, the transfer function of IIR notch filter can be obtained as

$$H_{NF}(z)=\frac{1}{2}\left[1+H_{allpass}(z)\right], \tag{1}$$

where $H_{allpass}(z)$ is the transfer function of optical allpass filter (OAF). For a 2N-order OAF, the transfer function can be defined as

$$H_{allpass}(z)=\frac{a_{2N}+\ldots+a_1 z^{-2N+1}+z^{-2N}}{1+a_1 z^{-1}+\ldots a_{2N} z^{-2N}}, \tag{2}$$

where a_1, …, a_{2N} are filter coefficients of OAF respectively. Allpass filters are pole-zero filters with unity magnitude response.

Gires-Tournois etalon (GTE) is a kind of allpass filter. According to the principle shown in Figure 1, notch filters can be accomplished by MGTI structure shown in Figure 2. It is formed by a total reflection reflector M and a multi-cavity GTE, a 50:50 beam splitter (BS) to output one group of channels, and an optical circulator to output another group of channels. GTE consist of $n + 1$ equally spaced reflectors. Reflectance and reflection coefficients of the $n + 1$ reflectors are 1, R_1, R_2, …, R_n, and 1, r_1, r_2, ... r_n, respectively. The spacing between $n + 1$ reflectors are d_1, d_2, ..., d_n respectively. The distance between M to BS is equal to that of the GTE to BS and is expressed as L.

The transfer function of the MGTI can be calculated as follows:

$$H(z) = \frac{1}{2}\left[1 + H_{GT}(z)\right], \tag{3}$$

where $H_{GT}(z)$ is the transfer function of multi-cavity GTE. Assume $z^{-1} = e^{-j\omega} = e^{-j4\pi d/\lambda}$, $d = C/2\Delta f$, where C is the speed of light in vacuum and Δf is frequency spacing of spectrum (FSR), the transfer function of GTE with $n + 1$ reflectors can be deduced from the recurrence formula as [10]:

$$H_{GT}(z) = H_{n+1}(z) = \frac{-r_n z + H_n(z)}{z - r_n H_n(z)};\ H_1(z) = 1, \tag{4}$$

where r_n is the reflection coefficients of the $n + 1$th reflector. H_n(z) is the transfer function of the GTE with n reflectors. It can be seen that the expression of Equation (3) is in accordance with that of Equation (1), so the proposed MGTI structure can get a notch response.

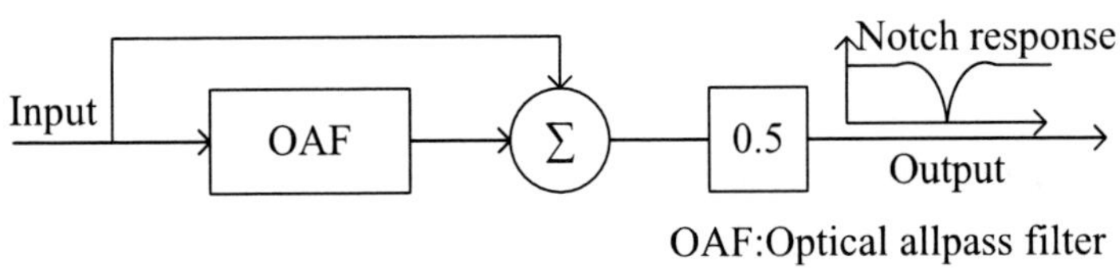

Figure 1. Principle of IIR optical notch filter based on optical allpass filter.

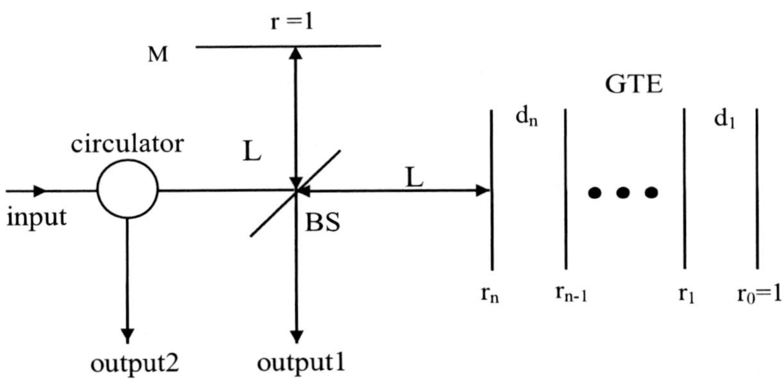

Figure 2. The structure of optical notch filter based on MGTI.

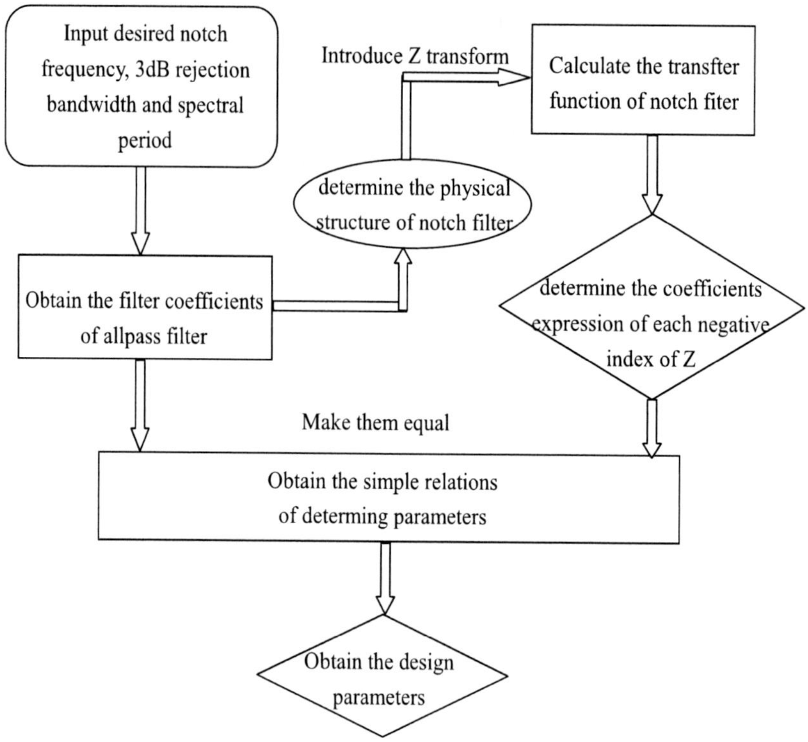

Figure 3. Solution flow chart of optical parameters.

Design Procedure

The whole process to get the design parameters of the notch filters can be seen in Figure 3.

We can summarize the entire design procedure of notch filter as follows:

1. obtain the filter coefficients of OAF according to the values of the desired notch frequency, 3 dB rejection bandwidth and spectral period (The filter coefficients of OAF can be obtained by using the method described in [13]).
2. determine the number of reflectors in GTE. Based on the above obtained filter coefficients of OAF, we can obtain the number of the filter coefficients of OAF. The number of the filter coefficients of OAF plus one equals that of reflectors in GTE. For OAF with cascaded multi-reflectors structure, the number of reflectors can also be decided by the numbers of notch point [14, 15]. If the notch filter to be designed has N notch points, the number of reflectors to be cascaded in GTE equals 2N + 1.
3. according to Equation (3) and (4), calculate the transfer function of notch filter determined in the second step. Thus the coefficients expression of each negative index of z can be known.
4. obtain the simple relations of determining the design parameters of notch filter by making the coefficients expression (obtained in the third step) equal to the filter coefficients of OAF (obtained in the first step). Finally, the design parameters of notch filter can be easily obtained by solving the relations.

Design Examples

As an example, a notch filter with single notch point in one FSR is designed. The location of the notch point is λ = 1550.3 nm, the 3 dB rejection bandwidth is $\Delta\lambda$ = 0.008 nm and free spectral range is FSR = 0.8 nm.

The transfer function of the optical notch filter in frequency domain after calculating is:

$$H_{NF}(z) = \frac{1}{2}\left(1 + \frac{0.9379 - 1.4983z^{-1} + z^{-2}}{1 - 1.4983z^{-1} + 0.9379z^{-2}}\right). \quad (5)$$

Equation (5) shows that there are two filter coefficients, so GTE with three reflectors should be used to design the notch filter. The cavity length $d_1 = d_2 = d = C/2\Delta f = 1.5\ mm$. The transfer function of the notch filter is deduced by Equation (3) and (4) as follows:

$$H(z) = \frac{1}{2}\left(1 + \frac{-r_2 - r_1 z^{-1} + r_2 r_1 z^{-1} + z^{-2}}{1 + r_2 r_1 z^{-1} - r_1 z^{-1} - r_2 z^{-2}}\right). \quad (6)$$

Comparing the Equation (5) and (6), we can obtain that Equation (5) equals Equation (6) when there exist the relations:

$$r_1(1 - r_2) = 1.4983;\ r_2 = -0.9379. \quad (7)$$

The design parameters of notch filter then can be easily obtained by solving the Equation (7). The obtained design parameters are r_1 = 0.7732 and r_2 = -0.9379, i.e., R_1 = 0.5978 and R_2 = 0.8797. The magnitude response of the filter is shown in Figure 4.

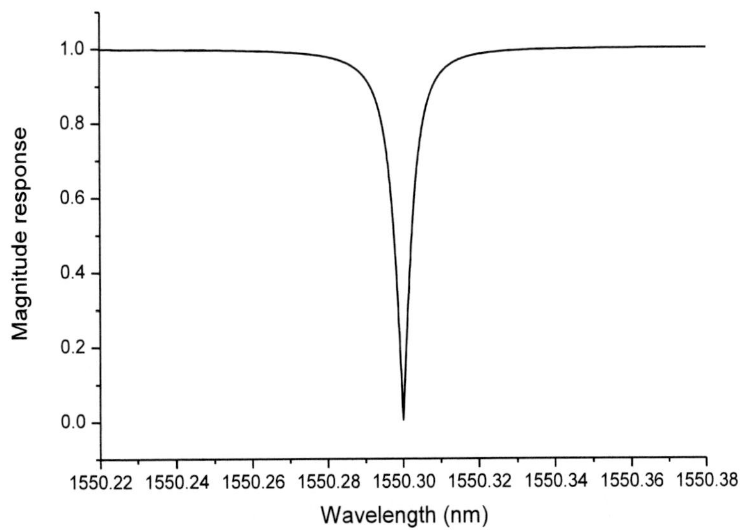

Figure 4. The magnitude response of designed notch filter with single notch point.

Multiple point Notch filters can also be designed using the method in a similar manner. Firstly, calculate the filter coefficients of OAF according to the values of the notch frequency, 3 dB rejection bandwidth and spectral period of the desired multiple point notch filter. Secondly, determine the number of reflectors in GTE, and calculate the transfer function of the multiple point notch filter by Equation (3) and (4), thus the coefficients expression of the transfer function of the multiple point notch filter can be known. Thirdly, by making the obtained coefficients expression of the transfer function of multiple point notch filter equal to the filter coefficients of OAF, the simple relations of determining the design parameters of the multiple point notch filter can be obtained. Finally, the design parameters of multiple point notch filter, i.e., the reflectance of reflectors can be easily obtained by solving the relations. Figure 5 (a) and 5 (b) are the magnitude response of the two notch point notch filters with equal and unequal 3 dB rejection bandwidth respectively. The detailed spectral characteristics and design parameters of Figure 5 are shown in Table 5.

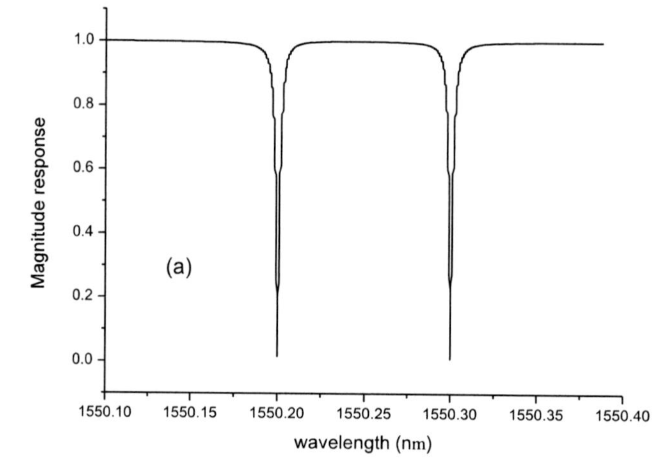

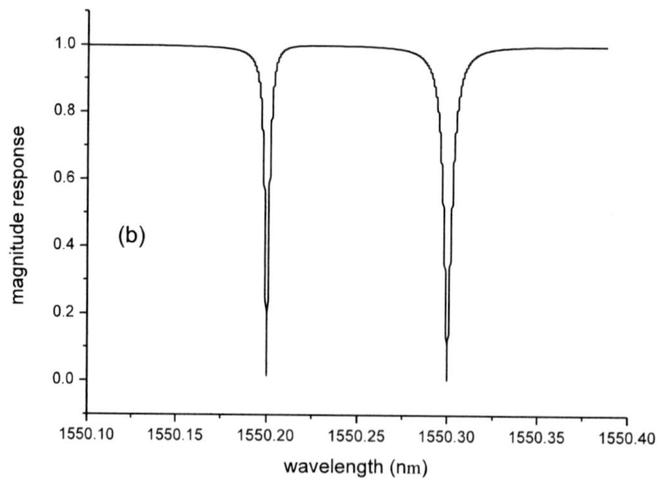

Figure 5. The magnitude response of designed two notch points notch filter with (a) equal 3 dB rejection bandwidth (b) unequal 3 dB rejection bandwidth.

The Design Parameters Analysis

In practical applications, the structural parameters can not always be accurate as the designed value. Hereinafter for simplicity taking the design example of single notch point notch filter above as an example, we will discuss the change of output spectrum when the reflectance and the spacing of reflectors shift from the designed value. The shift can be divided into two parts. One part is the shift of the reflectance of the reflectors, another part is the shift of the spacing between each reflectors.

1. The Spectra of Notch Filter When the Reflectance of GTE Has a Shift

Figure 6 (a) and Figure 6 (b) are the spectra of optical notch filter when R_1 and R_2 shift respectively.

From Figure 6, it can be seen that the R_1 and R_2 mainly control the notch location and 3 dB rejection bandwidth respectively. The notch point shifts to the long wave direction with the increasing of R_1, and the 3 dB rejection bandwidth deceases with the increasing of R_2. Table 2 shows the notch location and 3 dB rejection bandwidth of the filter when reflectance of the reflectors shift respectively.

Table 1. The spectral characteristics and design parameters of notch filters shown in Figure 5

Figure	Notch frequencies	3 dB rejection bandwidth	Design parameters
5 (a)	λ_1 = 1550.2 nm; λ_2 = 1550.3 nm	$\Delta\lambda_1$ = 0.004 nm; $\Delta\lambda_2$ = 0.004 nm	R_1 = 0.34; R_2 = 0.52 R_3 = 0.19; R_4 = 0.88
5 (b)	λ_1 = 1550.2nm; λ_2= 1550.3 nm	$\Delta\lambda_1$ = 0.004 nm; $\Delta\lambda_2$ = 0.008 nm	R_1 = 0.23; R_2 = 0.50 R_3 = 0.29; R_4 = 0.82

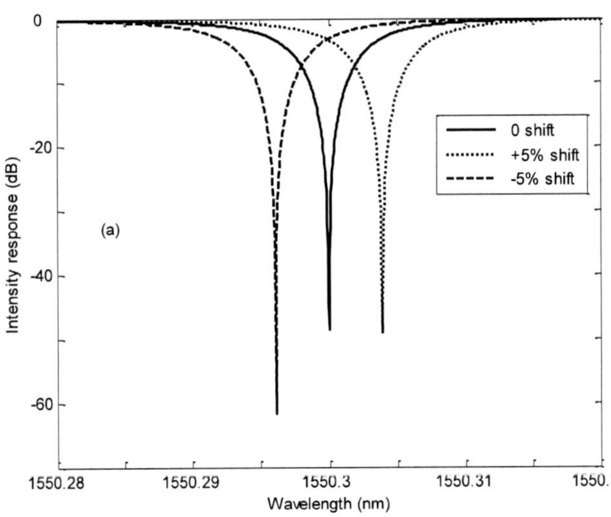

Figure 6. (Continued)

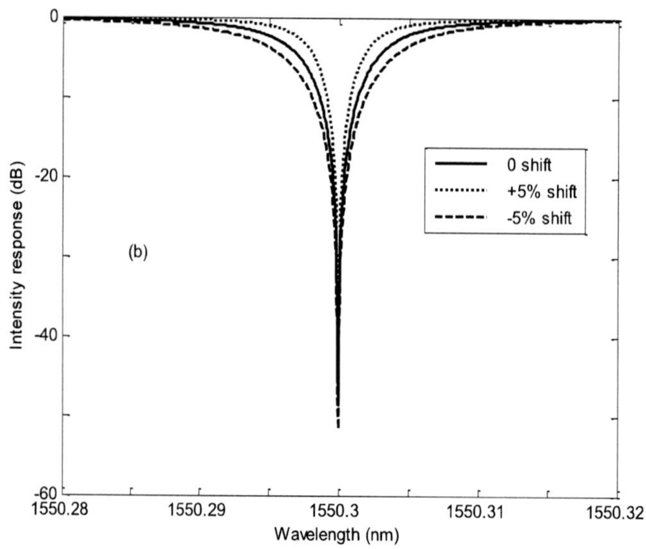

Figure 6. The intensity response of the notch filter (a) with R_1 shift; (b) with R_2 shift.

Table 2. The notch location and 3 dB bandwidth of the notch filter when R_1 and R_2 shift respectively

The shift of the reflectors' reflectance		Notch location	3 dB rejection bandwidth
The shift of R_1	5%	1550.30394	0.00809
	- 5%	1550.29617	0.00809
The shift of R_2	5%	1550.30	0.00505
	- 5%	1550.30	0.01146

Figure 7 (a) and (b) gives the notch location and 3 dB rejection bandwidth as a function of the shift rate of reflectance R_1 and R_2 respectively. From Figure 7, it can also be seen that R_1 influence notch location with 3 dB bandwidth unchanged and R_2 influence 3 dB rejection bandwidth with notch location unchanged. Furthermore, the relations of notch location to the shift rate of R_1 and 3 dB bandwidth to the shift rate of R_2 are all linear.

2. The Spectra of Notch Filter When the Spacing of GTE Has a Shift

In order to discuss conveniently, let D_1 and D_2 denote the values of cavity spacing after shift respectively, and $\delta d_1 = D_1 - d_1$, $\delta d_2 = D_2 - d_2$ denote the amount of the shift of the cavity spacing. Figure 8 (a) shows the magnitude response of the filter with small values of δd_1 and δd_2, and the magnitude responses of the filter with large values of δd_1 and δd_2 are shown in Figure 8(b). From Figure 8, we can see that when δd_1 and δd_2 are relatively small, the filter has the notch location tunability with the bandwidth nearly unchanged (see Figure 8(a)). And the location of notch point shifts to the long wave when the spacing is increasing. With the incessant increase of δd_1 and δd_2, the variation of bandwidth can't be neglected (see Figure 8(b)). Besides, we can also see that when spacing is smaller than the ideal value (such as curve F), the shift of the notch location is faster than that when spacing is larger than the ideal value with the same deviation amount (such as curve G).

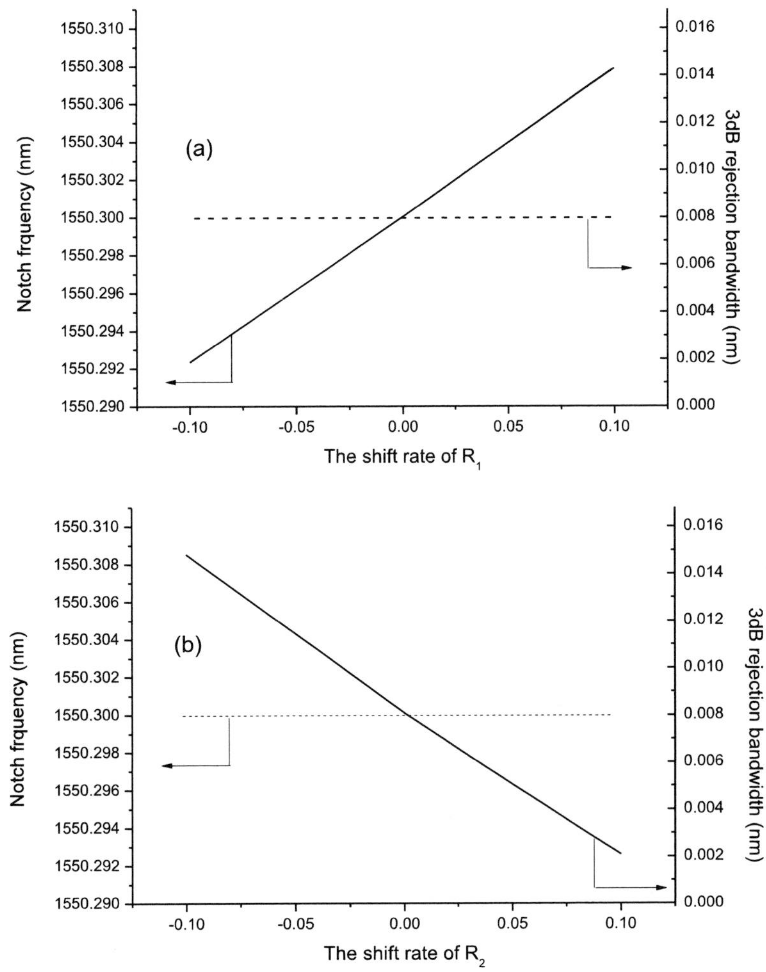

Figure 7. (a) The relation of notch location and 3 dB rejection bandwidth to the shift rate of R_1. (b) The relation of notch location and 3 dB rejection bandwidth to the shift rate of R_2.

Figure 9 gives the magnitude response of the filter with a special cavity length, i.e., $\delta d_1 + \delta d_2 = 0$. It can be seen that the filter has the characteristics of 3dB rejection bandwidth tunability when the values of cavity length of the two cavities are exchanged (see curve H and K; I and J).

Chromatic Dispersion Property

Chromatic dispersion (CD) is a key specification for optical filters. It greatly degrades the performance of the optical network systems. Figure 10 shows the CD of the single notch point notch filter in design example. We can see from Figure 10 that the notch filter has the high CD value in the range of near the notch location 1550.3 nm, while for other frequencies CD equal zero. So the notch filter has excellent CD property. It should be noted that in genearl CD is an unavoidable issue for etalon-based optical filters (for instance, interleavers).

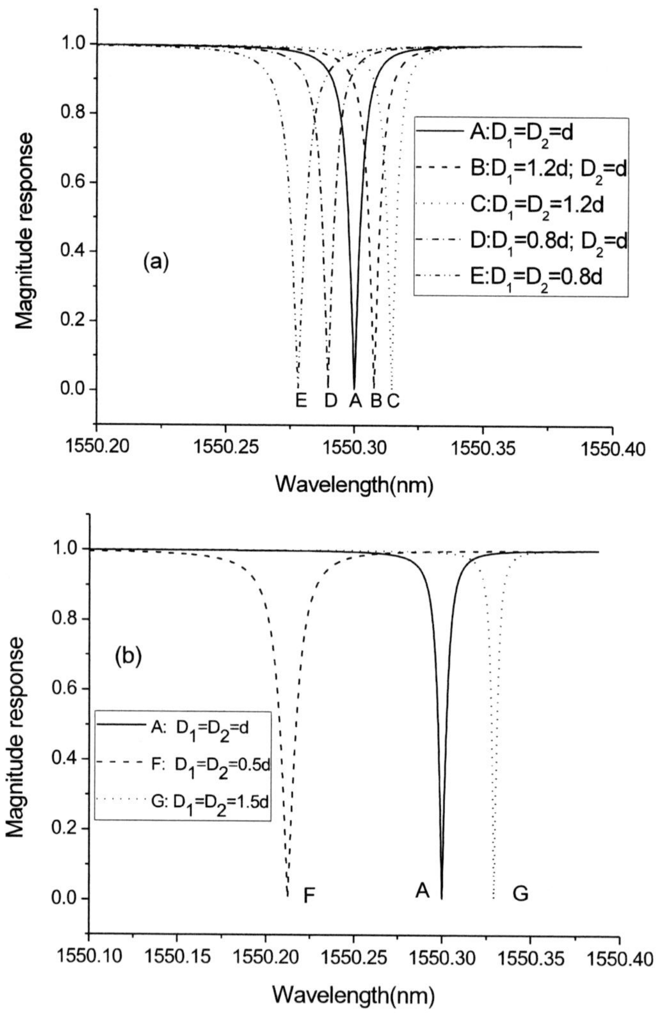

Figure 8. Magnitude response of filter (a) with small values of δd_1 and δd_2; (b) with large values of δd_1 and δd_2.

Independently Tunable Optical Notch Filter Based on Mach-Zehnder Interferometer with Double Ring Resonator Coupled Structure

In this section, from digital filter perspective using Z-tansform, a novel tunable optical notch filter is presented for Mach-Zehnder interferometer (MZI) with double ring resonator coupled structure [16]. The proposed design method to obtain the structural parameters of the notch filter is not only effective and simple, but also can be used to implement the optical notch filter with arbitrary notch frequency, 3 dB rejection bandwidth and notch point number. One design example of notch filter with two notch frequencies and different 3 dB rejection bandwidth is presented in detail. The change of the intensity response of the design example is also investigated for each design parameters deviating from ideal value. The results show that the notch frequency and 3 dB rejection bandwidth of each notch point can be tuned independently.

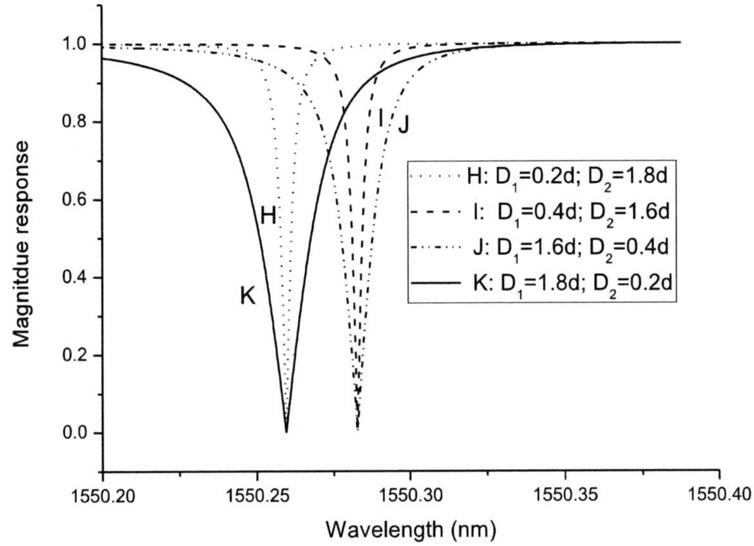

Figure 9. The magnitude response of the filter showing 3 dB rejection bandwidth tunability.

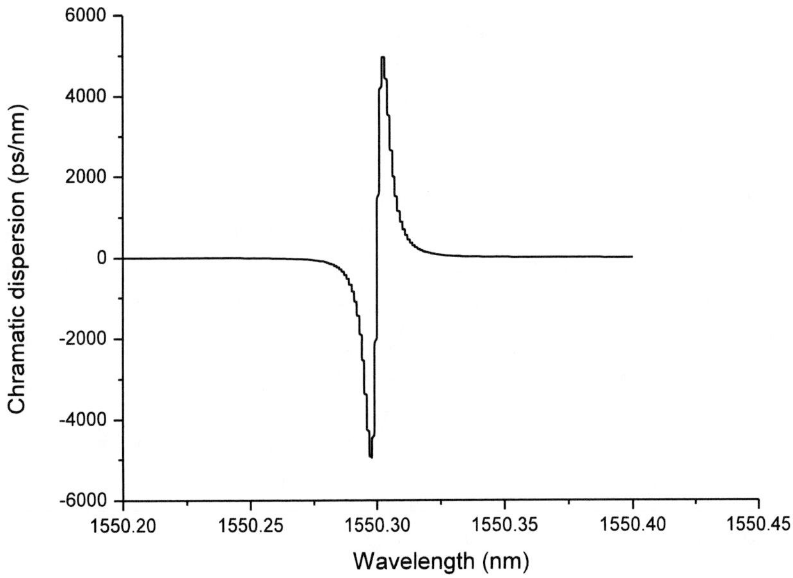

Figure 10. CD curve of the single notch point notch filter.

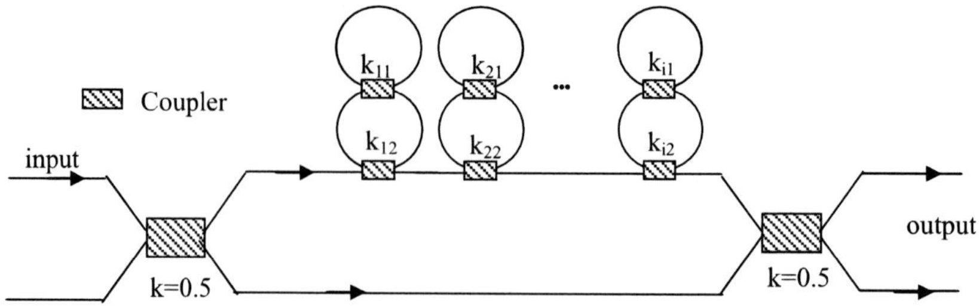

Figure 11. The physical structure to realize a notch filter.

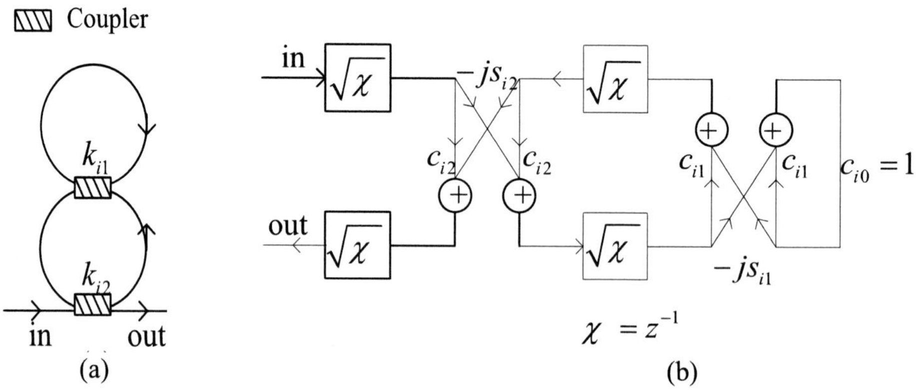

Figure 12. The (a) physical structure; (b) digital model of double ring resonator.

Design Principle

In the section before, we knew that the transfer function of notch filter can be obtained as Equation (1) and for a 2N-order allpass filter, the transfer function can be defined as Equation (2). Figure 11 is the structure of our proposed notch filter. In Figure 11, the optical allpass filter is composed of N cascaded double ring resonators. The length of up and down arm of the MZI is equal. The couplers at input and output ports of MZI have a power coupling ratio of 0.5. The power coupling ratios of the ith double ring resonator are k_{i1}, k_{i2} (i = 1, 2 .... N) respectively.

Figure 12(a) shows the structure of the ith double ring resonator. It's Z-transform representation is shown in Figure 12(b). Where $c_{i1}=\sqrt{1-k_{i1}}$, $c_{i2}=\sqrt{1-k_{i2}}$, $s_{i1}=\sqrt{k_{i1}}$, and $s_{i2}=\sqrt{k_{i2}}$, and k_{i1}, k_{i2} are the power coupling ratio of the first and the second coupler in the ith double ring resonator respectively.

From the Figure 12 (b), the transfer function of the ith double ring resonator can be deduced as follows:

$$H_{dou}^{i}(z)=\frac{c_{i2}-c_{i1}\left(c_{i2}+1\right)z^{-1}+z^{-2}}{1-c_{i1}\left(1+c_{i2}\right)z^{-1}+c_{i2}z^{-2}}, \tag{8}$$

where $z^{-1} = e^{-j2\pi fLn/C}$, f is the frequency of light, C is the speed of light in the vacuum, n and L are the refractive index and length of ring resonator respectively. The transfer function of the N cascaded double ring resonators can be expressed as

$$H_{allpass}(z) = H_{dou}^{1}(z) \cdot H_{dou}^{2}(z) \ldots \cdot H_{dou}^{N}(z) \tag{9}$$

The transfer matrix of the coupler at the input and output of MZI can be expressed as

$$\Phi_{cplr}^{in} = \Phi_{cplr}^{out} = \begin{bmatrix} \sqrt{1-k} & -j\sqrt{k} \\ -j\sqrt{k} & \sqrt{1-k} \end{bmatrix} \tag{10}$$

where $k = 0.5$ is the power coupling ratio of the coupler. The delay matrix is written for both arms as follows

$$\Phi_{delay} = \begin{bmatrix} H_{allpass}(z) & 0 \\ 0 & 1 \end{bmatrix}. \tag{11}$$

According to the theory of transfer matrix, the overall transfer matrix for the filter is the product of matrices

$$H(z) = \Phi_{cplr}^{out} \cdot \Phi_{delay} \cdot \Phi_{cplr}^{in} \tag{12}$$

From Equation (8) to Equation (12), we can derive that the transfer function of notch filter in Figure 11 is given by

$$H(z) = \frac{1}{2}[1 + \prod_{i=1}^{N} \frac{c_{i2} - c_{i1}(c_{i2}+1)z^{-1} + z^{-2}}{1 - c_{i1}(1+c_{i2})z^{-1} + c_{i2}z^{-2}}] \tag{13}$$

It can be seen that the Equation (13) is in accordance with Equation (1), so the filter of this structure can get a notch response.

Design Method

In order to obtain notch response with different spectral characteristics, the number of cascaded double ring resonators and the power coupling ratios of all the double ring resonators need to be determined. According to the spectral characteristics of the desired optical notch filter, we can calculate the notch filter's transfer function in frequency domain shown in Equation (2) utilizing the method described in Reference [13]. Thus the order of the notch filter can be known. On this basis, the number of cascaded double ring resonator can be determined. It is equal to the half of the order of the notch filter. Then the optical structure of the notch filter can be determined and its transfer function can be obtained as Equation (13).

The power coupling ratios of all the double ring resonators can be obtained by making Equaion (13) equal to Equation (2).

In signal processing technology when the transfer function of a filter is known, it is usually straightforward to obtain all the values of poles of the filter. Because $H_{allpass}(z)$, the transfer function of allpass filter in Equation (2), is a function of z with real coefficients, its poles/zeros are symmetric about the real axis. If one of the poles of $H_{allpass}(z)$ is expressed as $z_i = r_i e^{j\theta_i}$, r_i and θ_i are the modulus and argument of z_i respectively, $z_i^* = r_i e^{-j\theta_i}$ must be another pole of $H_{allpass}(z)$. Therefore the denominator of $H_{allpass}(z)$ must include the item as follows:

$$(1-z_i z^{-1})(1-z_i^* z^{-1}) = 1-2r_i\cos\theta_i z^{-1} + (r_i)^2 z^{-2} . \tag{14}$$

Let Equation (14) equal to the denominator of $H_{dou}^{i}(z)$, we obtain

$$\begin{cases} c_{i2} = (r_i)^2 \\ c_{i1} = \dfrac{2r_i\cos\theta_i}{1+(r_i)^2} \end{cases} . \tag{15}$$

The above is the case for single pole. The power coupling ratios of all the double ring resonators can be obtained similarly. The proposed method to obtain the design parameters of double ring resonator is much simpler than directly solving the Equation (2) and Equation (13) simultaneously, especially for the case that have many poles.

Design Example

As an example, a notch filter with two notch points in one spectral period is designed. The locations of two notch points are λ_1 = 1550.2 nm, λ_2 = 1550.3 nm, and the 3 dB rejection bandwidth of two notch points are $\Delta\lambda_1$ = 0.008 nm, $\Delta\lambda_2$ = 0.016 nm respectively. Spectral period (FSR) equals 0.8 nm. Firstly, according to the desired spectral characteristics and the method described in Reference [13], the filter coefficients of allpasss filter can be obtained as follows: a_1 = -1.6491, a_2 = 2.1008, a_3 = -1.5287 and a_4 = 0.8211. Therefore the transfer function of the filter to be designed in frequency domain equals:

$$H(z) = \frac{1}{2}\left[1+\frac{0.8211-1.5287z^{-1}+2.1008z^{-2}-1.6491z^{-3}+z^{-4}}{1-1.6491z^{-1}+2.1008z^{-2}-1.5287z^{-3}+0.8211z^{-4}}\right] . \tag{16}$$

From Equation (16), we know that the order of the allpass filter equals 4. So two double ring resonators are needed to construct such an optical notch filter. The transfer function of the notch filter can be calculated from Equation (13) as follows:

$$H(z)=\frac{1}{2}[1+\frac{c_{12}-c_{11}\left(c_{12}+1\right)z^{-1}+z^{-2}}{1-c_{11}\left(1+c_{12}\right)z^{-1}+c_{12}z^{-2}}\cdot\frac{c_{22}-c_{21}\left(c_{22}+1\right)z^{-1}+z^{-2}}{1-c_{21}\left(1+c_{22}\right)z^{-1}+c_{22}z^{-2}}]. \tag{17}$$

Figure 13 is the pole/zero plot of allpass filter shown in Equation (17). We can obtain that the values of the poles are $p_{1,2} = 0.09915617 \pm 0.9610582j$ and $p_{3,4} = 0.7253938 \pm 0.5944996j$ respectively. $j=\sqrt{-1}$. The design parameters can then be easily obtained using Equation (15) as $c_{12} = 0.9335$, $c_{11} = 0.1026$, $c_{22} = 0.8796$, $c_{21} = 0.7719$. The intensity response of the filter corresponding to these structural parameters is shown in Figure 14. From Figure 14, it can be seen that the design result is in good agreement with the target filter.

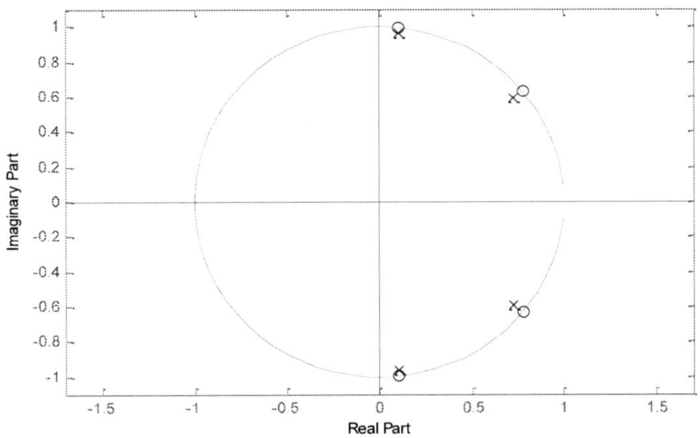

Figure 13. Pole/zero plot.

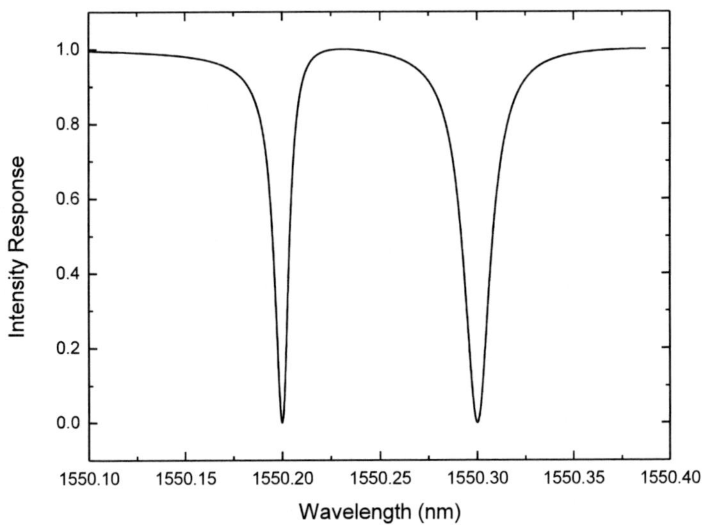

Figure 14. The intensity response of notch filter with unequal 3 dB rejection bandwidth.

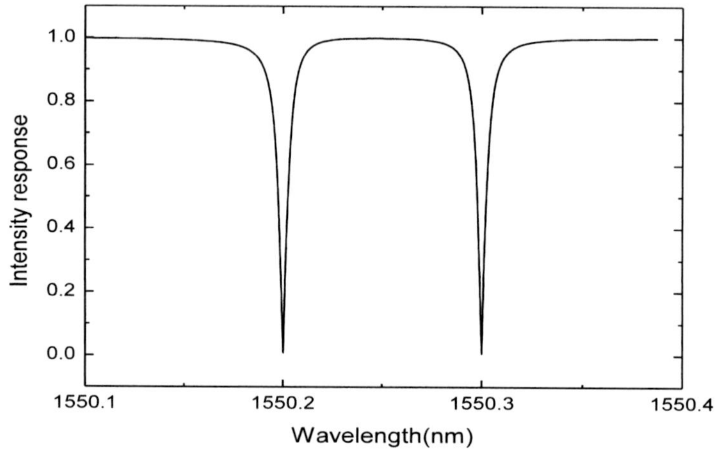

Figure 15. The intensity response of notch filter with equal 3 dB rejection bandwidth.

Optical notch filters with other spectral characteristics can also be designed using the method similarly. Figure 15 shows the intensity response of the notch filter with equal 3 dB rejection bandwidth. When only one double ring resonator is used in the structure, the notch filter with single notch point can be realized.

Independently Tunable Characteristics

The tunable characteristics of notch filter can be obtained from analyzing the design parameters. Hereinafter taking the design example above as an example, we will discuss the change of output spectrum when the design parameters shift from the designed value. Because power coupling ratios are often used in practical applications, design parameters $c_{12} = 0.9335$, $c_{11} = 0.1026$, $c_{22} = 0.8796$, $c_{21} = 0.7719$ are transformed into power coupling ratios as $k_{12} = 0.1286$, $k_{11} = 0.9895$, $k_{22} = 0.2263$, and $k_{21} = 0.4042$.

Figure 16 (a-d) show the intensity spectral of the filter with k_{11}, k_{12}, k_{21} and k_{22} shift respectively. From the figure we can see that parameter k_{12}, k_{11} determine the characteristics of the first notch point (1550.2 nm), and parameter k_{22}, k_{21} determine the characteristics of the second notch point (1550.3 nm). That is to say, each double ring resonator introduces a notch at its resonance frequency. If single double ring resonator, such as the first double ring resonator is only used in the structure, single notch point notch filter with notch location at 1550.2 nm would be obtained. The intensity response is shown in Figure 17 (a). Similarly, when the second double ring resonator is only used in the structure, the notch filter with notch location at 1550.3 nm would be obtained. Its intensity response is shown in Figure 17 (b). The 3 dB rejection bandwidth of the first and the second notch point is controlled by k_{12} and k_{22} respectively and the 3 dB rejection bandwidth increase with the increasing of the k_{12} and k_{22}. The notch location of the first and second notch point is determined by k_{11} and k_{21} respectively and with the decreasing of k_{11} and k_{21}, the notch location of notch filter shifts to the long wavelength.

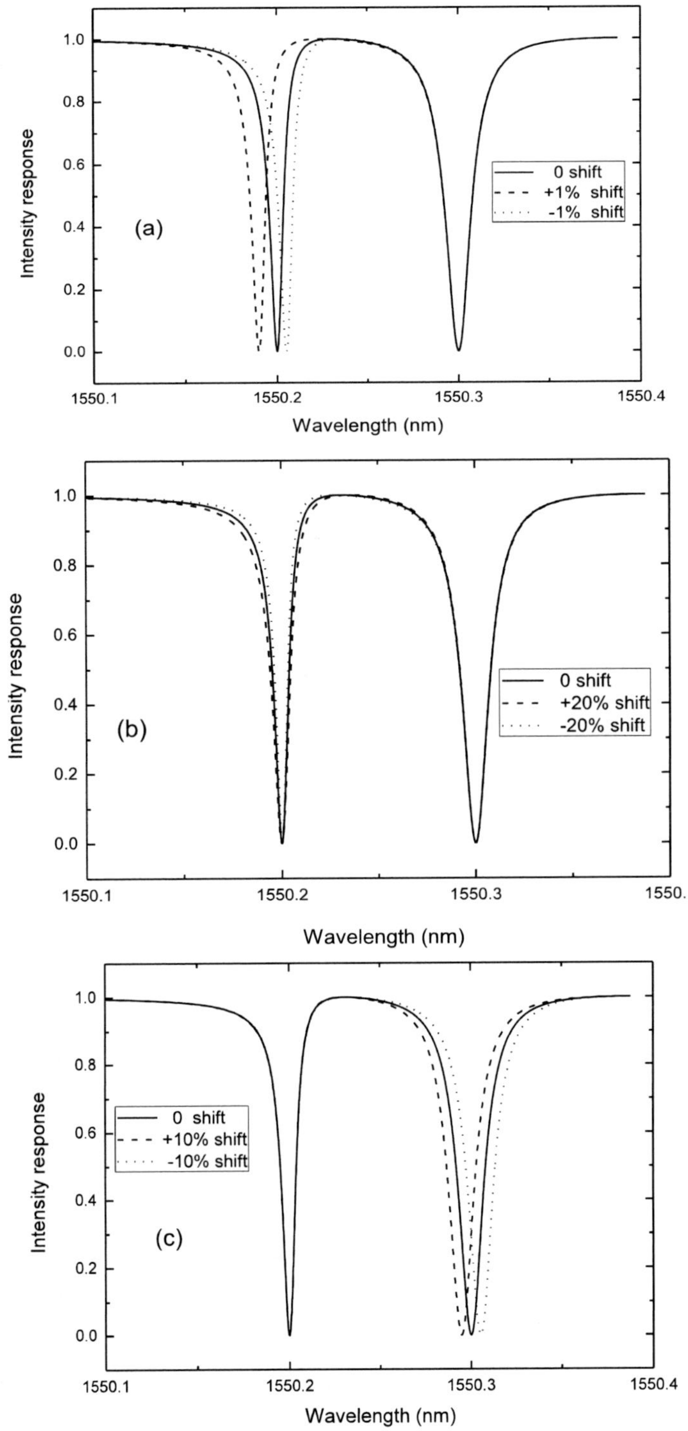

Figure 16. (Continued)

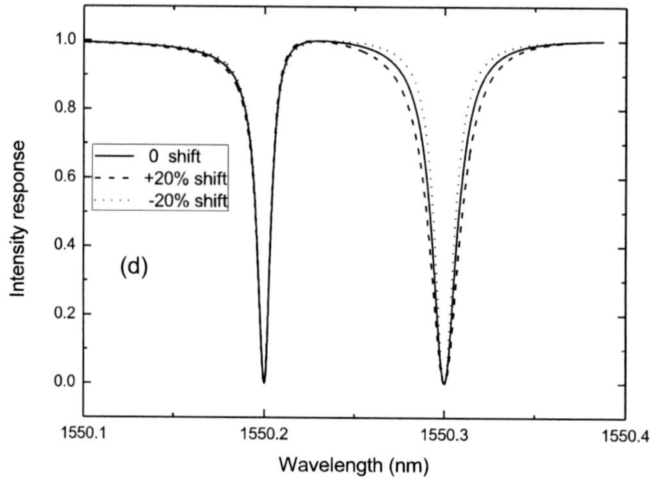

Figure 16. The notch response when parameters (a) k_{11}; (b) k_{12} (c) k_{21}; (d) k_{22} shift respectively.

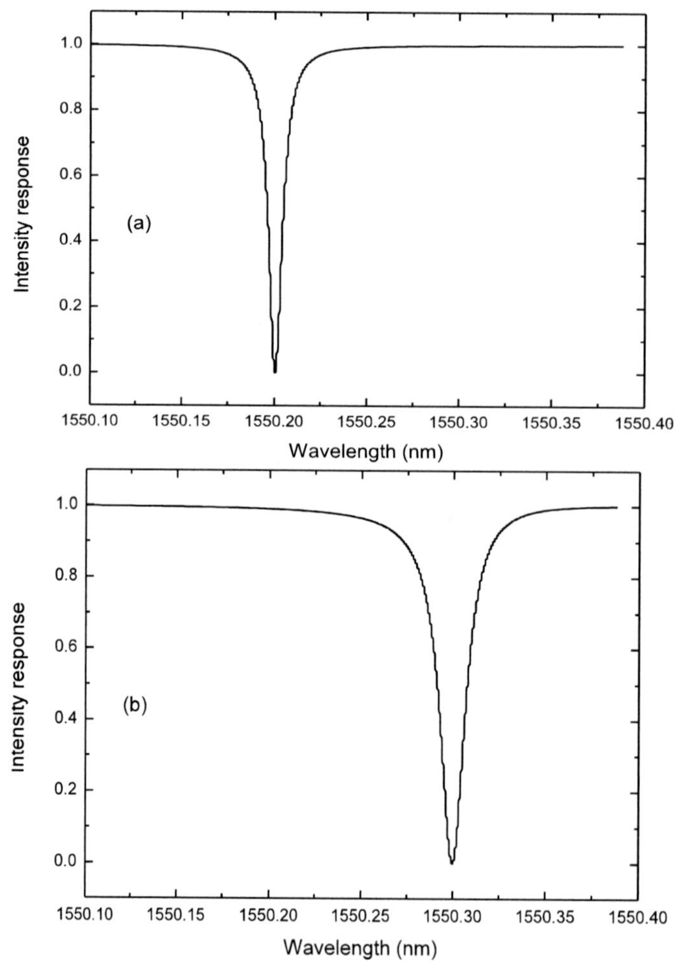

Figure 17. The intensity response of notch filter with notch location at (a) 1550.2 nm; (b) 1550.3 nm.

Comparing the Figure 16 (a-d), it can be seen that the filter is most sensitive to k_{11}, and then are the k_{21}, k_{22}, and k_{12} in sequence. Furthermore, when k_{11} is larger than the ideal value, the change of notch location of the first notch point is faster than that when k_{11} is smaller than the ideal value. So in practical applications, we would rather k_{11} smaller than larger. While for other three parameters, i.e., k_{21}, k_{22}, and k_{12}, the relations between the change of spectral characteristics and the shift of parameters are linear.

From the analysis above, it can be seen that the notch frequency and 3 dB rejection bandwidth of each notch point can be independently tuned. The tunability of notch frequency and 3 dB rejection bandwidth of the *i*th notch point can be realized through adjusting parameters k_{i1} and k_{i2} respectively.

REFERENCES

[1] Alessio Vallese, Andrea Bevilacqua, Analysis and design of an integrated notch filter for the rejection of interference in UWB Systems, *IEEE Journal of Solid-State Circuits*, vol. 44, pp. 331-343, 2009.

[2] C.D. McManus, D. Neubert, E. Cramer, Characterization and elimination of AC noise in electrocardiograms: a comparison of digital filtering methods, *Computers and Biomedical Research*, vol. 26, pp. 48-67, 1993.

[3] D.V.B. Rao, S.Y. Kung, Adaptive notch filtering for the retrieval of sinusoids in noise, *IEEE Transactions on Acoustics, Speech, and Signal Processing*, vol. 32, pp. 791-802, 1984.

[4] Soo-Chang Pei, Chien-Cheng Tseng, A novel structure for cascade form adapative notch filters, *Signal Processing*, vol. 33, pp. 95-110, 1993.

[5] M.S. Rasras, K.Y. Tu, D.M. Gill et al., Demonstration of a tunable microwavephotonic notch filter using low-loss silicon ring resonators, *J. Lightwave Technol.* 27 (2009) 2105-2110.

[6] S. Nikitin, C. Manka, J. Grun, Modifiedˇ Solc notch filter for deep ultraviolet applications, *Opt. Soc. Am.* 48 (2009) 1184-1189.

[7] N. You, R.A. Minasian, A novel tunable microwave optical notch filter, IEEE Trans. Microw. *Theory Tech.* 49 (2001) 2002-2005.

[8] J. Zhang, L. Liu, Novel Mach-Zehnder interferometer structure for tunable optical interleaver, *Optical Engineering*, vol. 45, pp. 045003, 2006.

[9] Z. P. Wang, S. J. Chang, C. Y. Ni, and Y. J. Chen, A high-performance ulrtacompact optical interleaver based on double-ring assisted Mach–Zehnder interferometer, *IEEE Photon. Technol. Lett.*, vol. 19, pp. 1072-1074, 2007.

[10] J. Zhang, X. Yang, Universal Michelson Gires-Tournois interferometer optical interleaver based on digital signal processing, *Optics Express*, vol. 18, pp. 5075-5088, 2010.

[11] Chao–Hsing Hsieh, Ruibo Wang, Zhiqing James Wen et al., Flat-top interleavers using two Gires-Tournois etalons as phase-dispersion mirrors in a Michelson interferometer, *IEEE Photon. Technol. Lett.*, vol. 15, pp. 242-244, 2003.

[12] J. Zhang, S. Guo, X. Li, General IIR optical notch filter based on Michelson Gires-Tournois interferometer, *Optics Communications.* vol. 285, pp. 491-496, 2012.

[13] Soo-Chang Pei, Chien-Cheng Tseng, IIR multiple notch filter design based on allpass filter, *IEEE Trans. Circuits Syst. II: Analog Digital Signal Process*. vol. 44, pp. 133-136, 1997.
[14] C.K Madsen, Jian H. Zhao, Optical filter design and analysis-a signal processing approach, John wiley&Sons, Inc., New York, 1999.
[15] James A. Cadzow, Digital notch filter design procedure, *IEEE Trans. Acoustics, Speech and Signal Processing*, vol. 22, pp. 10-15, 1974.
[16] J. Zhang, S. Guo, X. Li, Independently tunable optical notch filter based on double ring resonator structure, *Optik*, vol. 124, pp. 1307-1310, 2013.

In: Digital Signal Processing (DSP)
Editor: Juan Zhang

ISBN: 978-1-63485-168-8

Chapter 9

DESIGN AND IMPLEMENTATION OF LOW-PASS, HIGH-PASS AND BAND-PASS FINITE IMPULSE RESPONSE (FIR) FILTERS USING FPGA

Warsame H. Ali* and Emmanuel S. Kolawole

Department of Electrical and Computer Engineering,
Prairie View A and M University, Prairie View, TX, US

ABSTRACT

This chapter presents the design and implementation of a Low-pass, a High-pass and a Band-pass Finite Impulse Response (FIR) Filter using SPARTAN-6 Field Programmable Gate Array (FPGA) device. The Filter performance is tested using Filter design and Analysis (FDA) and FIR tools from Mathworks. The FDATool is used to define the Filter order and coefficients, and the FIR tool is used for Simulink simulation. The FPGA implementation is carried out using Spartan-6 LX75T-3FGG676C for different Filter specifications and simulated with the help of Xilinx ISE (Integrated Software Environment). System Generator ISE design suit 14.6i is used in synthesizing and co-simulation for FPGA Filter output verification. Finally, comparison is done between the results obtained from the software simulations and those from FPGA using hardware co-simulation. The simulation waveforms and synthesis reports verify the parallel implementation of FPGA which proves its effectiveness in terms of speed, resource usage and power consumption.

Keywords: digital filters, FIR Filter, matlab simulink, FDATool, FIR tool, distributed srithmetic, FPGA, xilinx system generator

* Corresponding author: Warsame H. Ali. E-mail: whali@pvamu.edu.

Introduction

In signal processing, the function of a filter is to remove unwanted parts of the signal, such as random noise, or to extract useful parts of the signal, such as the components lying within a certain frequency range [1]. The FIR Filter block diagram is as shown in Figure 1 [10].

An ideal filter is a network that allows signals of only certain frequencies to pass while blocking all others. Depending on the region of frequencies that are allowed through or not, filters are characterized as low-pass, high-pass, band-pass, band-reject and all-pass. There are many needs for electric filters, some of the more common being those used in radio and television sets, which allow tuning into a certain channel by passing its band of frequencies while filtering out those of other channels [2].

The FIR Filters are widely used in signal processing and can be implemented using programmable digital processors. Due to the high performance requirements and increasing complexity of DSP and multimedia communication applications, filters with large number of taps are required to increase the performance in terms of high sampling rate. As a result the filtering operations are computationally intensive and more complex in terms of hardware requirements [3]. The FIR filters perform the weighted summations of input sequences with constant coefficients in most of the signal processing and multimedia applications. The FIR filter is a digital filter widely used in Digital Signal processing applications in various fields like imaging, instrumentation and communications. Programmable digital processor signal (PDPS) can be used in implementing the FIR filter. Nowadays, to make the difference on the market, new industrial control systems have to be highly performing, very flexible and reliable. At the same time, the cost is a key issue.

In order to reduce the cost, time-to-market has to be shortened, the price of a controller device has to be cheap and its energy consumption needs to be reduced [4]. This cost reduction is all the more challenging that new industrial control systems arebased on ever increasing sophisticated control algorithms which need a lot of computing resources and need reduced execution time. However, in realizing a large order filter many complex computations are needed which invariably affects the general performance of the common digital signal processors in terms of speed, cost flexibility, stability and so on.

Methods

The methods used in this chapter are based on FDATool Filter Design, FIR Compiler and Simulink Model, System Generator/Hardware Co-Simulation and FIR Filter Implementation on FPGA.

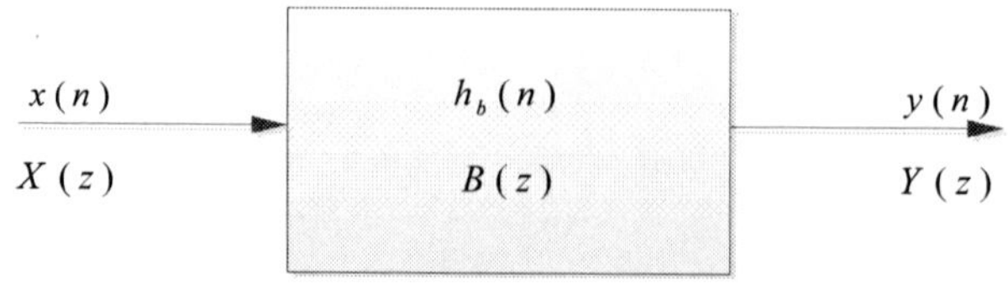

Figure 1. A block diagram of a FIR filter [10].

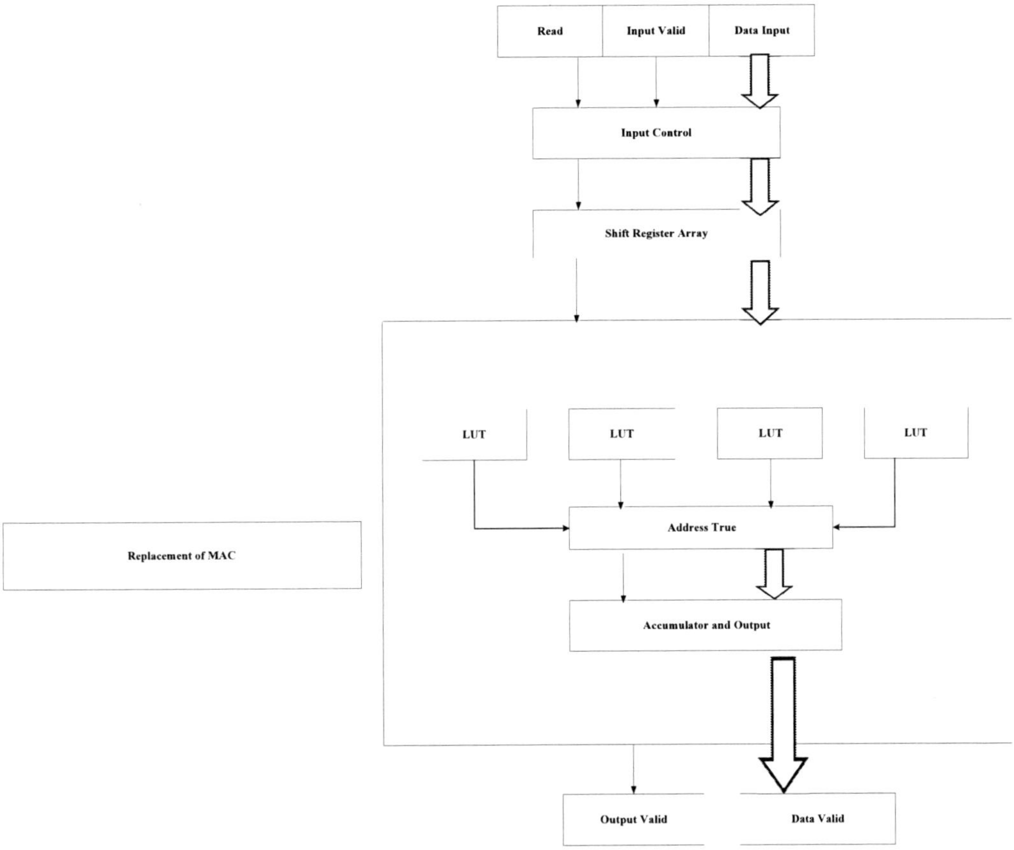

Figure 2. Implementation of FIR Filter using DA [6].

In this paper FDATool from Matlab mathematical computational package with digital signal processing toolboxes is used to design filter response and generate coefficients tables. In the proposed approach, FIR tool utilizes distributed arithmetic (DA) as shown in Figure 2 which actually uses lookup table for storing constant coefficients. So the use of lookup tables reduces the hardware complexity and hence the new design is more efficient in terms of less area, more speed and low power consumption due to its parallel implementation on FPGA, unlike the traditional DSP that utilizes MAC (Unit multiplier and add accumulator) which increases memory resources as Filter order increases [6]. System Generator package provided by Xilinx is used for FPGA implementation of each Filter specification. The overall Filter output waveforms and synthesis reports performances under parallel implementation of FPGA is greatly enhanced with the proposed method.

FIR Filters Design

The design of an FIR filter for a specific application includes calculating the coefficients according to different criteria like filter order, sampling frequency, pass-band and stop-band frequencies, etc. The coefficient calculating could be done by different software which makes it as simple task. There are different methods used in designing Digital FIR Filters, such as

Equiripple, window, least-square, frequency sampling and interpolated FIR method. In this paper, I have chosen Equiripple method because:

- It meets specifications with the least number of coefficients
- Uses less amount of resources on FPGA for implementation
- The weighted approximation error between the desired and actual frequency response is spread evenly across the passband and stopband of the Filter thereby minimizing error
- Passband and stopband deviations can be specified separately

Below is the Equirriple equation;

$$\max \varepsilon(j2\pi f) = f \in [-0.5, 0.5] \left| Q(j2\pi f)[H_d(j2\pi f) - H(e^{j2\pi f})] \right| \quad (1)$$

where:

$H(e^{j2\pi f})$ is the best approximation frequency response
$H_d(j2\pi f)$ is the ideal frequency response
Q() is the weighting function
$\varepsilon(j2\pi f)$ is the Equiripple factor

Table 1 as shown below elaborates the specifications of Low-pass, High-pass and Band-pass Filters used in this paper;

- Fs is the sampling frequency.
- F-stop is the stopband frequency.
- F-pass is the passband frequency.
- A-pass is the passband attenuation.
- A-stop is the stopband attenuation.

Table 1. FIR Filters specifications

Options	Low-Pass Filter	High-Pass Filter	Band-Pass Filter
Design Method	FIR Equiripple	FIR Equiripple	FIR Equiripple
Frequency Specifications	Units: KHz Units: kHz Fs: 1500 F-stop: 300 F-pass: 270	Units: kHz Fs: 1500 F-stop: 450 F-pass: 480	Units: kHz Fs: 1500 F-stop 1: 270 F-stop 2: 480 F-pass 1: 300 F-pass 2: 450
Magnitude specification	Units: dB A-stop: 54 dB A-pass: 1 dB	Units: dB A-stop: 54 dB A-pass: 1 dB	Units: dB A-stop 1: 54 dB A-stop 2: 54 dB A-pass: 1 dB

FIRs have the advantage of being much more realizable in hardware [7] because they avoid division and feedback paths. FIR filter response y(n) is computed by operation between filter coefficients b(k) and the data input x(n) as shown by the equation:

$$y[n]=\sum_{k=0}^{N-1} h_k.x[n-k] \tag{2}$$

where $x[n-k]$ = represents the filter input, h_k = represents the filter coefficients, $y[n]$ = represents the filter output, N = represents the number of filter coefficients (order of filter).

The value of the constant k is a minimum value for which the expression $N \le 2^k$ is valid.

$$k = \lfloor 1 + log_{(2)}N \rfloor.$$

where the operator $\lfloor \quad \rfloor$ represents rounding down to a less value.

In FIR Filter design, Filter frequency response coefficients and the corresponding window type function must be known before Filter hardware realization. Table 2 shows the transfer function equations used in different types of Filters [8].

The FIR Filter is graphically represented by a direct form approach [7] as shown in Figure 3. This graphical Filter design approach using Equiripple method enhances easy selections for specifying passband, filter order, and design methods, as well as provides Filter response of individual specification. The FDATool from Mathworks as shown for Low-pass Filter in Figure 4 is used to generate MATLAB model design and coefficient tables based on specification. Following the same procedure, High-pass and Band-pass Filters are equally generated based on their specifications.

Table 2. The frequency responses of standard ideal Filters [8]

Type of Filter	Frequency Response $h_d[n]$
Low-pass Filter	$h_d[n]=\left(\dfrac{\sin\left[\omega_c(n-M)\right]}{\pi(n-M)}\right.$;n ≠ M $\dfrac{\omega_c}{\pi}$ $n=M$
High-pass Filter	$h_d[n]=\left(-\dfrac{\sin(\omega_c(n-M))}{\pi(n-M)}\right.$;$n=M$ $1-\dfrac{\omega_c}{\pi}$; n ≠ M
Band-pass Filter	$h_d[n]=\left(\dfrac{\sin(\omega_{c2}(n-M))}{\pi(n-M)}-\dfrac{\sin(\omega_{c1}(n-M))}{\pi(n-M)}\right.$;$n \neq M$ $\dfrac{\omega_{c2}-\omega_{c1}}{\pi}$; $n=M$

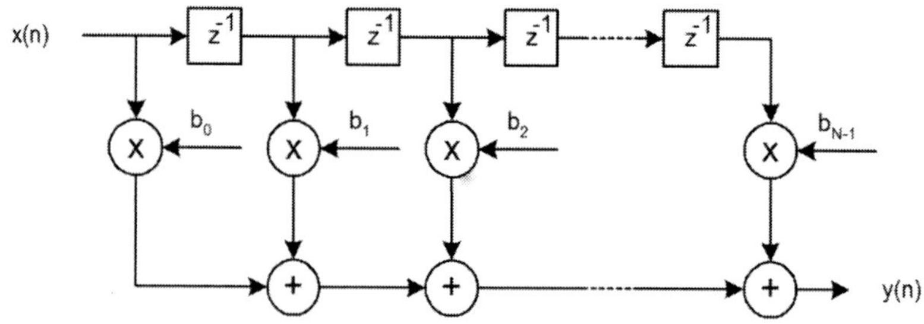

Figure 3. The direct architecture of a typical FIR Filter [7].

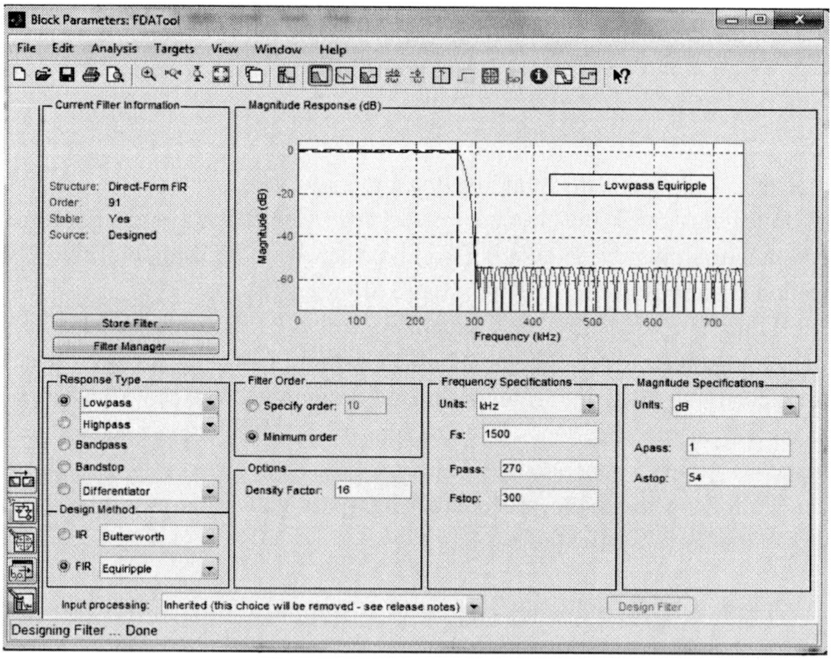

Figure 4. Low-pass Filter using FDATool.

FPGA Implementation of FIR Filter

After designing the filters based on their specifications from Matlab, the Xilinx software package provided by Spartan-6 FPGA board, System Generator is then used for the appropriate FIR FPGA Filter implementation for Low-pass, High-pass, Band-pass Filter as shown in Figures 5, 6 and 7.

Simulation Results for Low-Pass FIR Filter

Figure 5 displayed the FPGA implementation on Low-pass Filter. Figures 8b and 8c verified the comparison between the Low-pass Filter simulation from Matlab and FPGA implementation. Figure 8a is the Input signal below (F_C) using chirp source. Where (F_C) is the cut-off frequency.

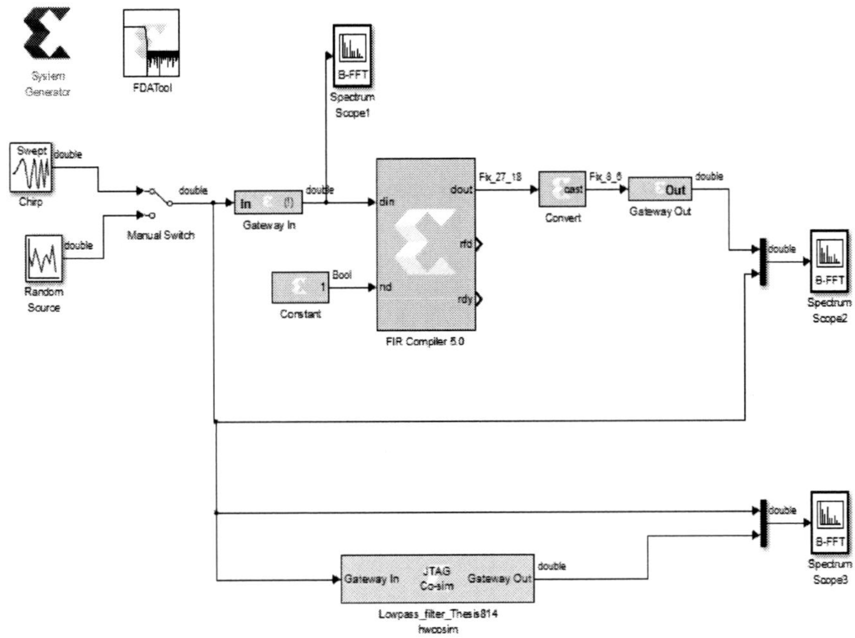

Figure 5. Low-pass FIR Filter.

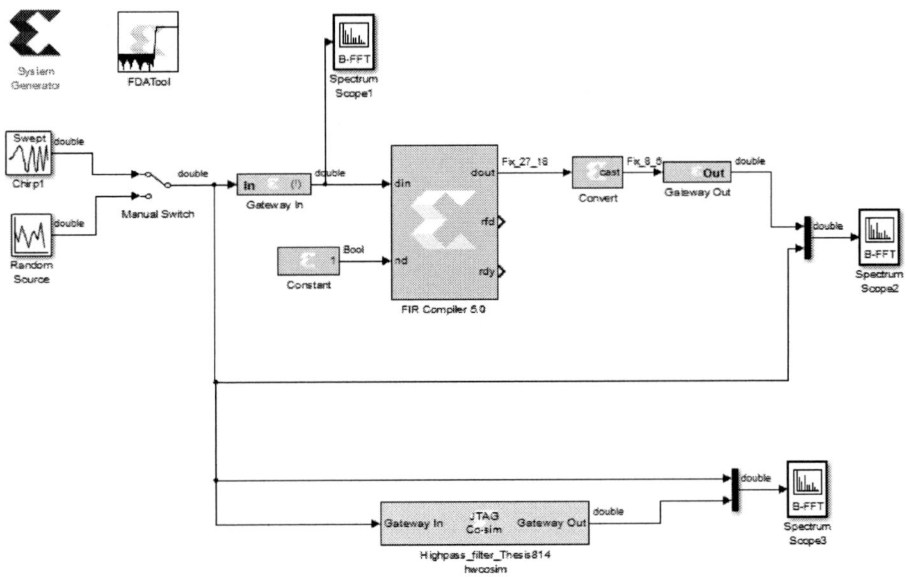

Figure 6. High-pass FIR Filter.

Figures 8d, 8e and 8f shown below analyzed when the input signal is above the cut-off frequency and no output waveforms for Low-pass frequency.

Figure 6 displayed the FPGA implementation on High-pass Filter while the corresponding Figures 9b and 9c verified the comparison between the High-pass Filter simulation from Matlab and FPGA implementation. Figure 9a is the input signal using chirp source when the signal is above the cut-off frequency.

Figures 9d, 9e and 9f as shown below represents the output waveforms from Matlab and FPGA for High-pass when the input signal is below (F_C). High-pass did not pass at low frequency.

Figure 7 displayed the FPGA implementation on Band-pass Filter while the corresponding Figures 10b and 10c verified the comparison between the Band-pass Filter simulation from Matlab and FPGA implementation. Figure 10a is the input signal using chirp source when the signals are between (F_{C1}) and (F_{C2}). Band-pass passes frequency when the frequency is within the specified bandwidth.

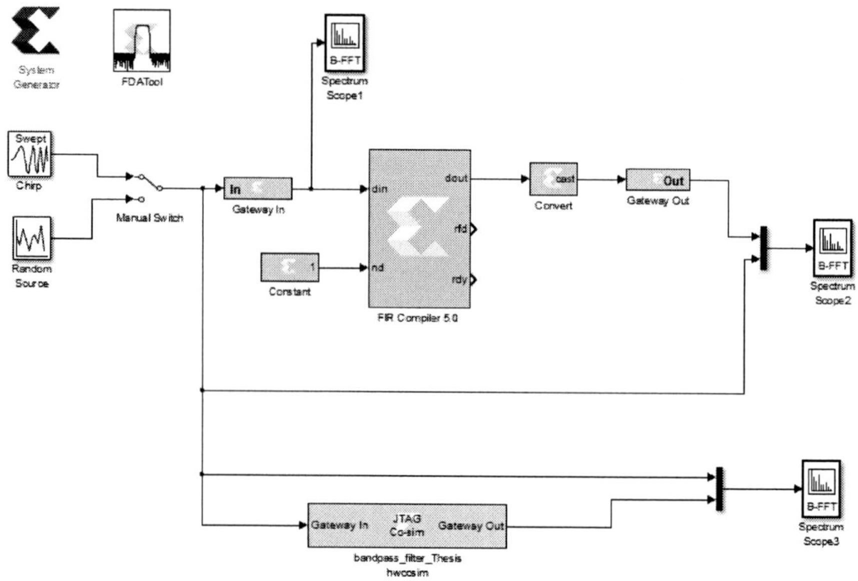

Figure 7. Band-pass FIR Filter.

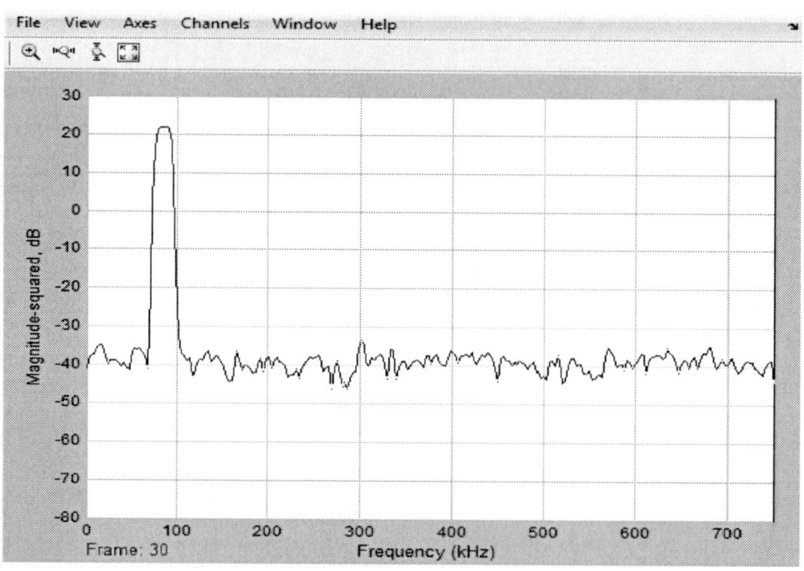

a

Figure 8. (Continued)

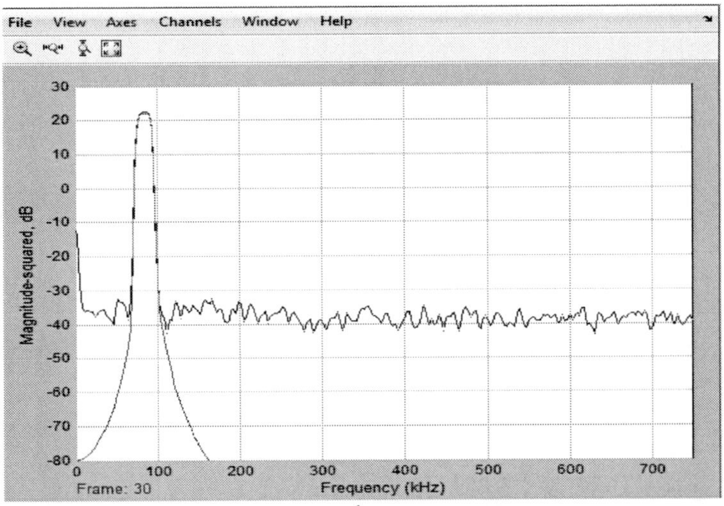

b

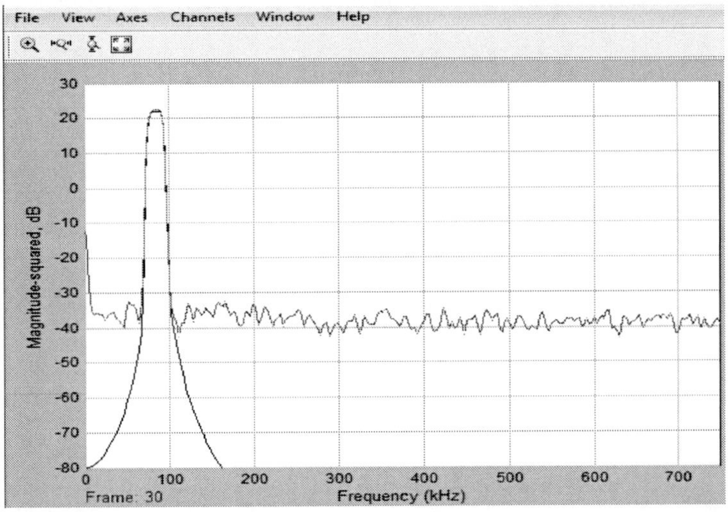

c

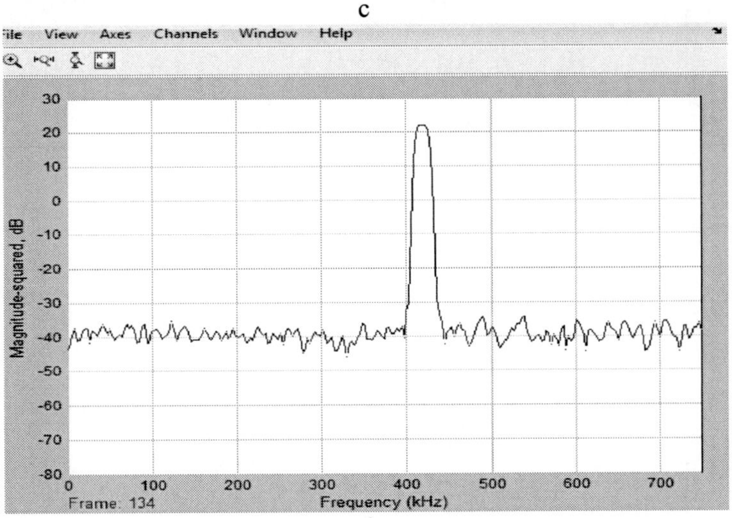

d

Figure 8. (Continued)

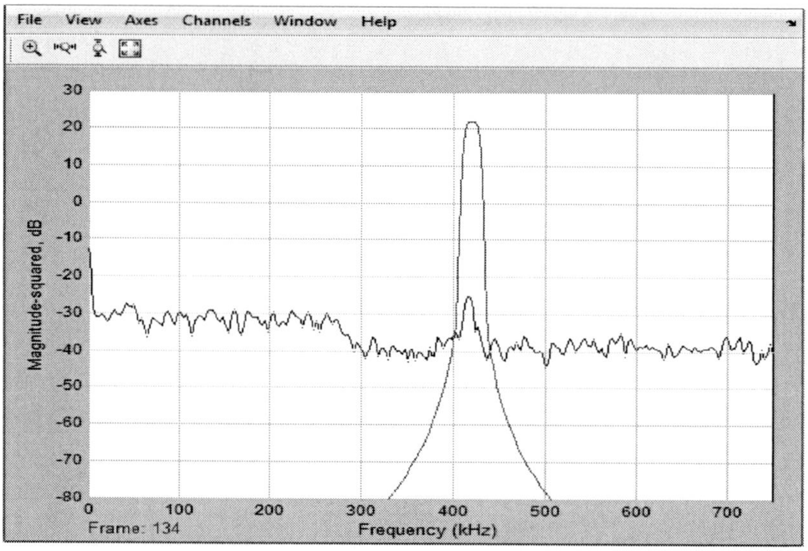

e

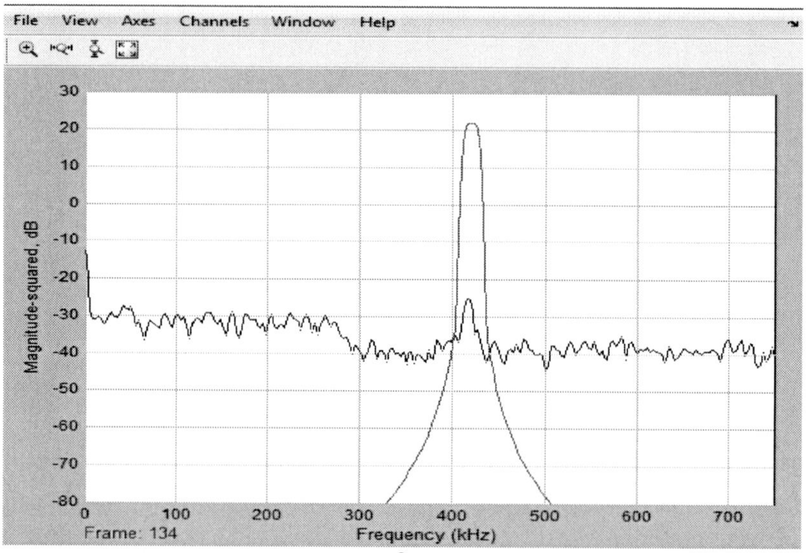

f

Figure 8. a. Input signal below $(\boldsymbol{F_C})$ using chirp source, b. Output waveform from Low-pass FIR filter using Matlab, c. Output waveform from Low-pass FIR filter using FPGA, d. Input signal above $(\boldsymbol{F_C})$ using chirp source, e. Output waveform from low-pass FIR filter using Matlab, f. Output waveform from low-pass filter using FPGA.

Figures 10d, 10e and 10f as shown below represents the output waveforms from Matlab and FPGA for Band-pass when the input signal is below (F_{C1}). Band-pass did not pass at frequency lower that (F_{C1}).

Figures 10g, 10h and 10i as shown below represents the output waveforms from Matlab and FPGA for Band-pass when the input signal is below (F_{C1}). Band-pass did not pass at frequency above that (F_{C2}).

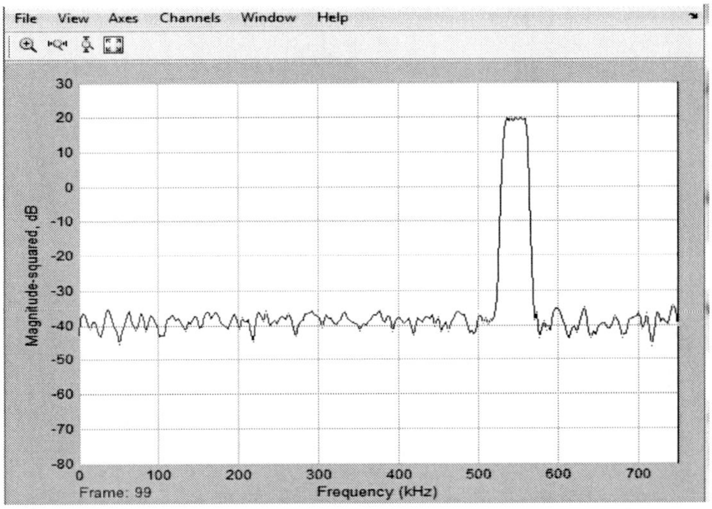

a

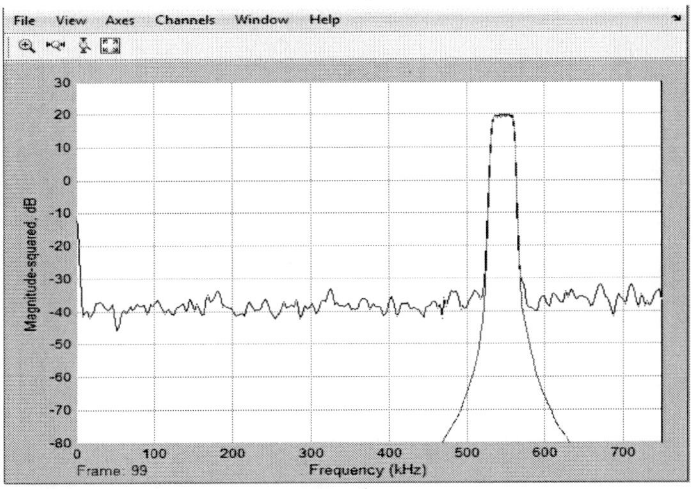

b

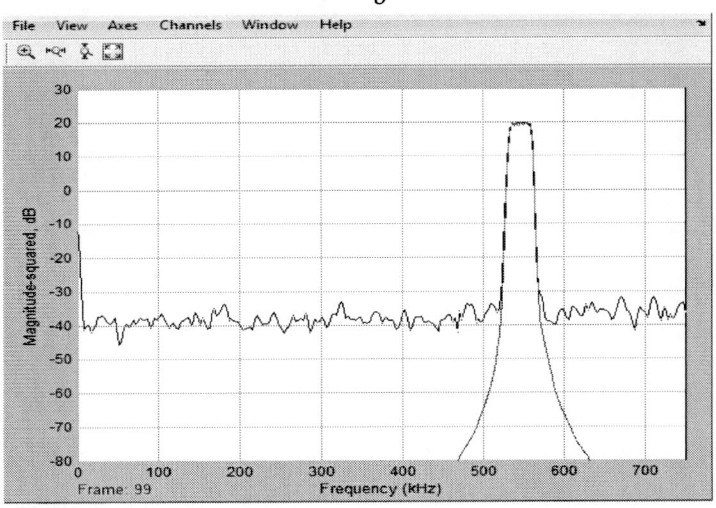

c

Figure 9. (Continued)

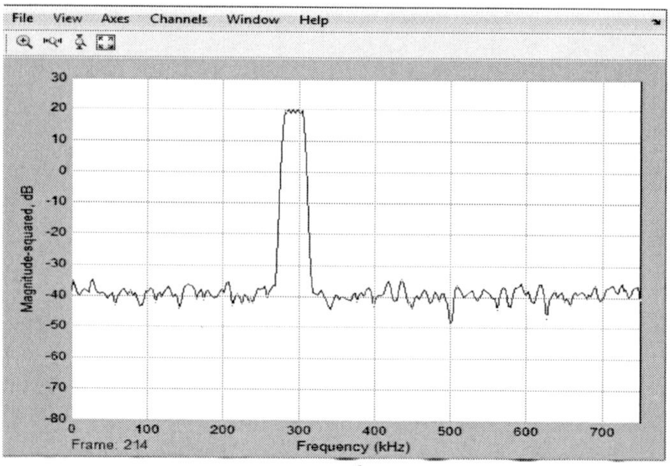

d

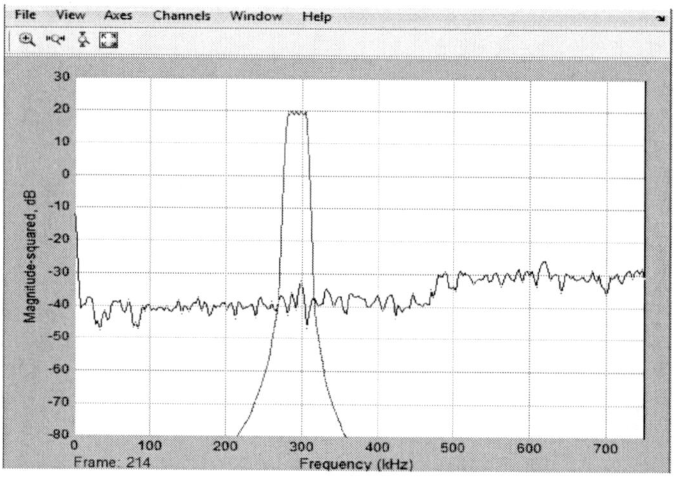

e

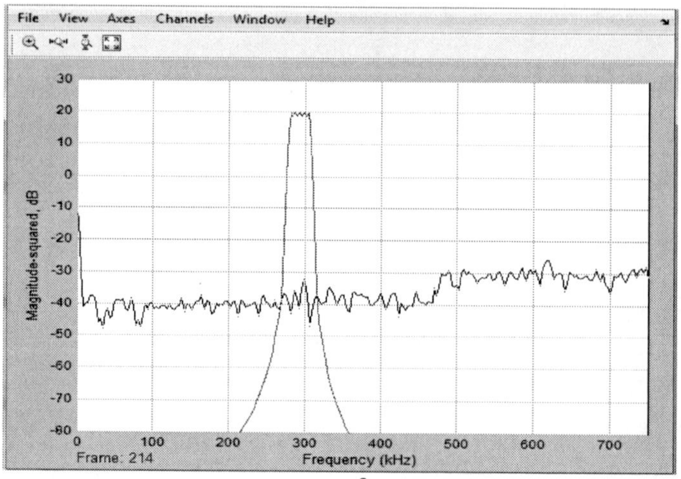

f

Figure 9. a. Input signal above ($\boldsymbol{F_C}$) using chirp source, b. Output waveform from High-pass FIR filterusing Matlab, c. Output waveform from High-pass FIR filterusing FPGA, d. Input signal below ($\boldsymbol{F_C}$) using chirp source, e. Output waveform from high-pass FIR filter using Matlab, f. Output waveform from high-pass filter using FPGA.

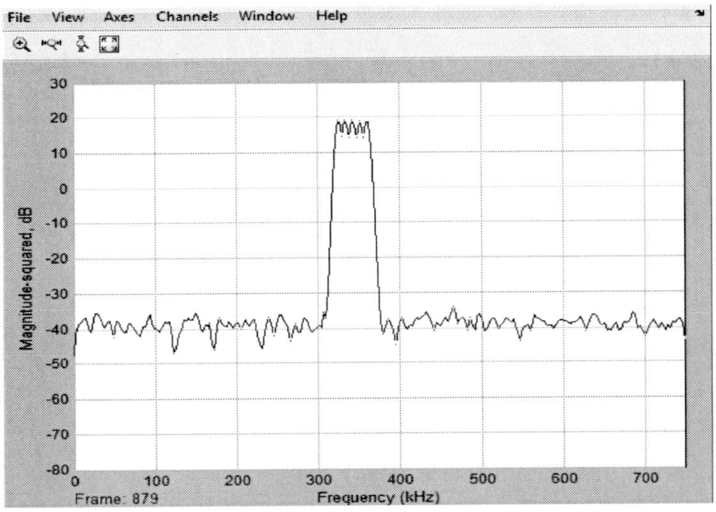

a

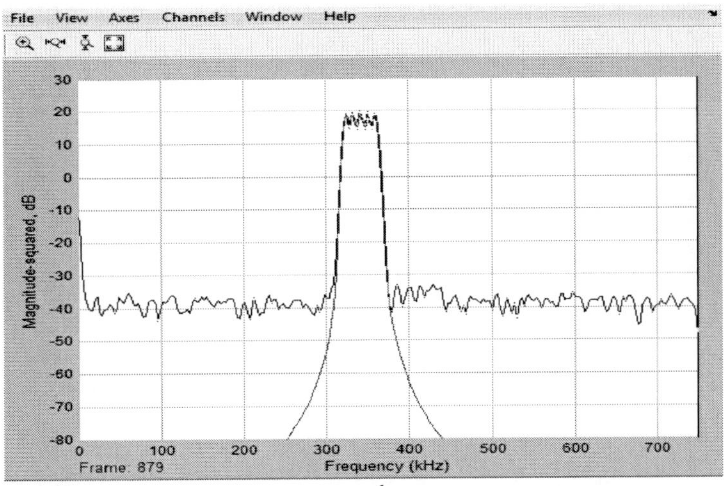

b

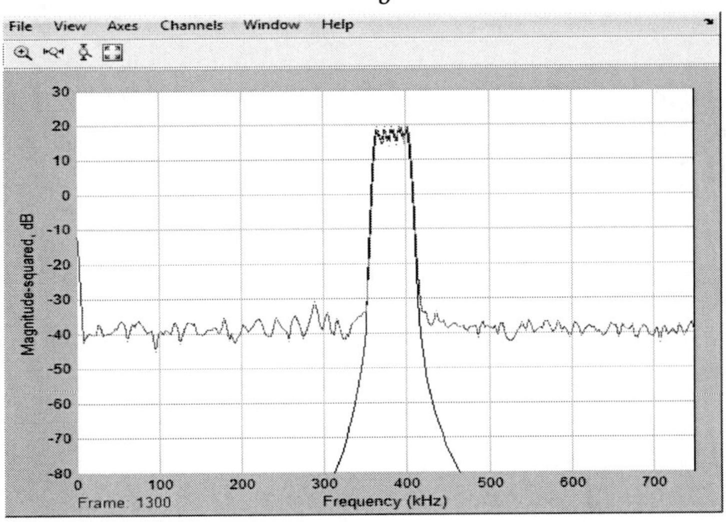

c

Figure 10. (Continued)

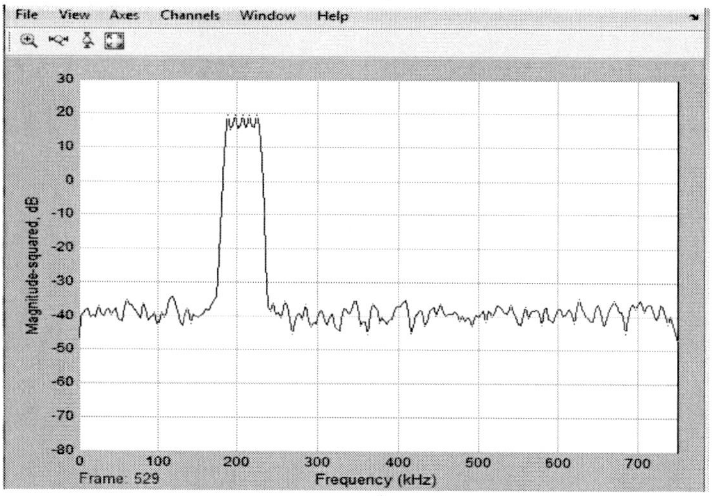

d

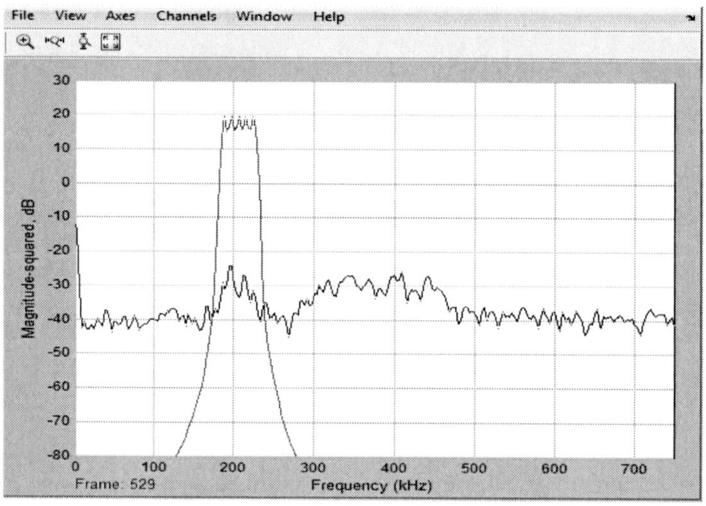

e

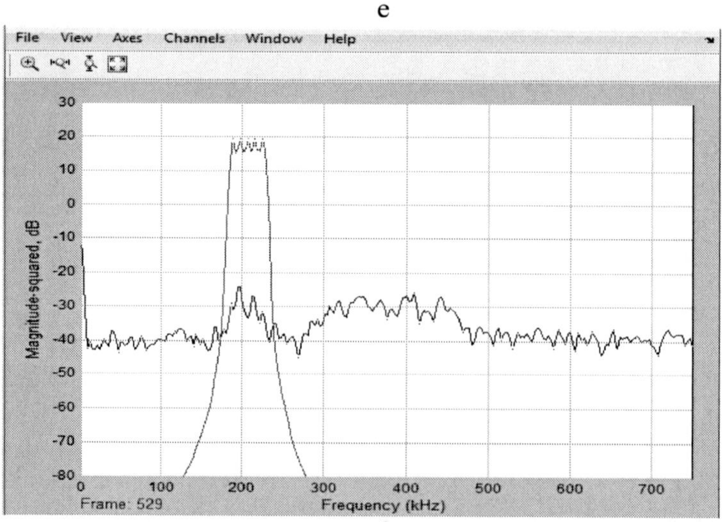

f

Figure 10. (Continued)

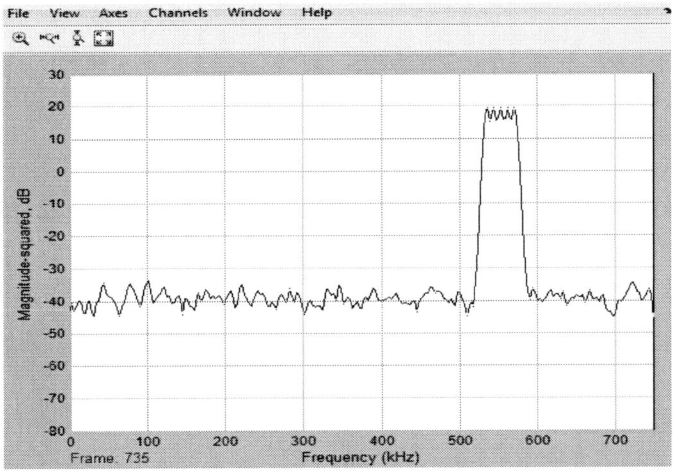

g

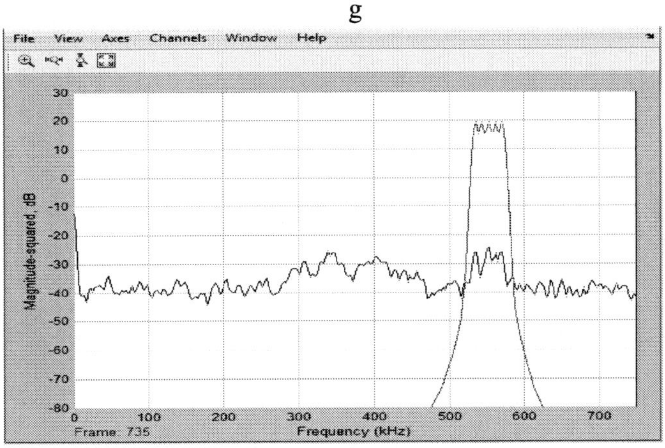

h

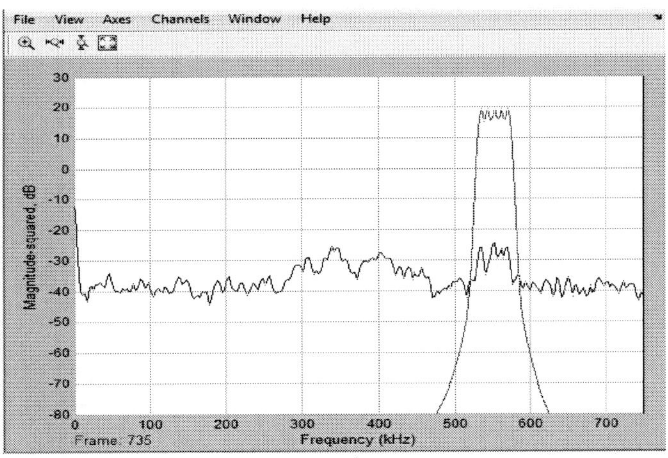

i

Figure 10. a. Input signal between $(\boldsymbol{F_{C1}})$ and $(\boldsymbol{F_{C2}})$ using chirp source, b. Output waveform from Band-pass FIR filter using Matlab, c. Output waveform from Band-pass FIR filter using FPGA, d. Input signal below $(\boldsymbol{F_{C1}})$ using chirp source, e. Output waveform from band-pass FIR filter using Matlab, f. Output waveform from band-pass filter using FPGA, g. Input signal above $(\boldsymbol{F_{C2}})$ using chirp source, h. Output waveform from band-pass FIR filter using Matlab, i. Output waveform from band-pass filter using FPGA.

Parallel Implementation on FPGA

Figure 11 as shown below identify the true parallelism nature of FPGA, so different processing operations do not have to compete for the same resources. Each independent processing task is assigned to a dedicated section of the chip, and can function autonomously without any influence from other logic blocks [9]. As a result, the performance of one part of the application is not affected when more processing blocks are added [3].

Simulation Results for Parallel Implementation

When the three filters were combined in parallel with individual specifications, the same output waveforms for all the filters were observed which actually proved the parallelism nature of FPGA. With input signal using Random source as shown in Figure 12a, Figures 12b, 12c, 12d, 12e, 12f and 12g verified the comparison between the Low-pass, High-pass and Band-pass Filters simulations from Matlab and FPGA parallel implementation.

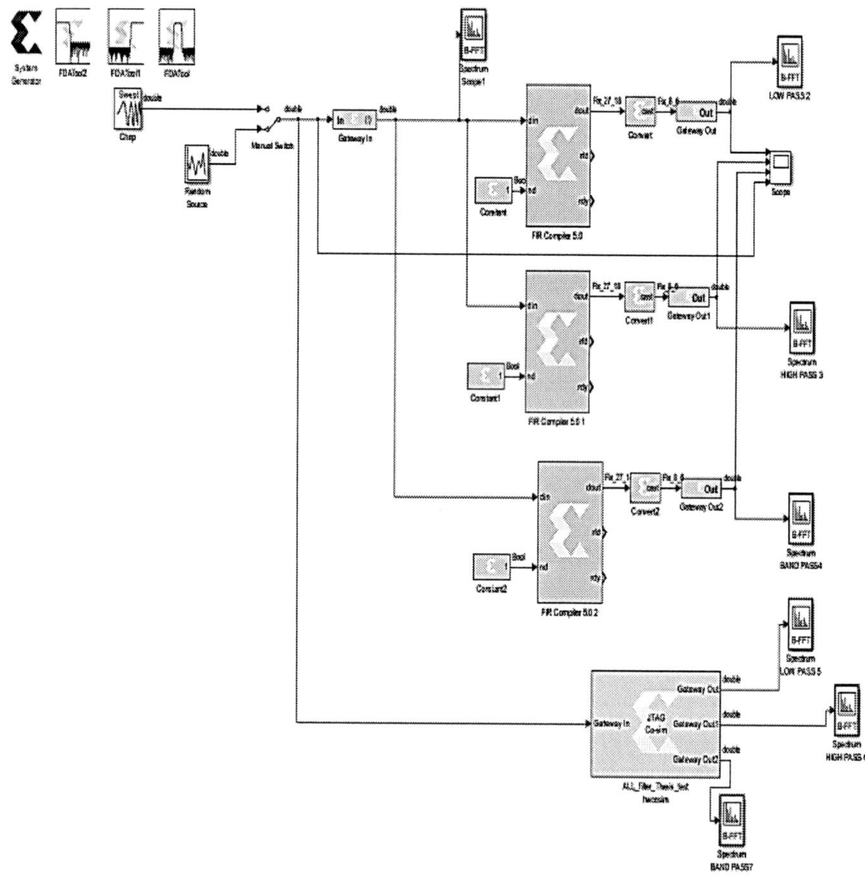

Figure 11. Parallel implementation of FIR Filters.

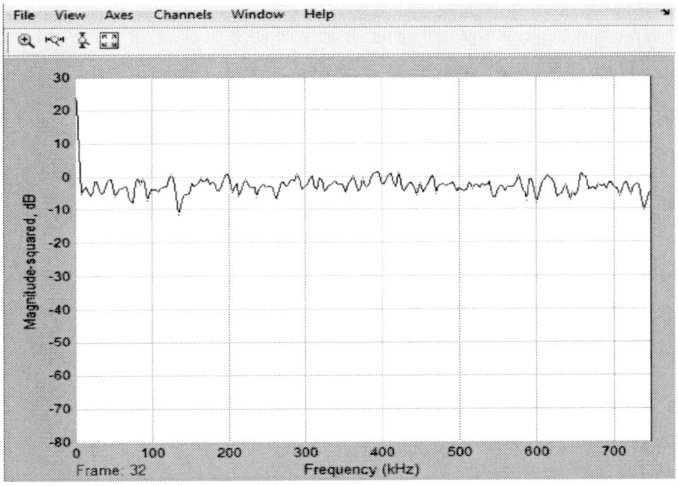

a

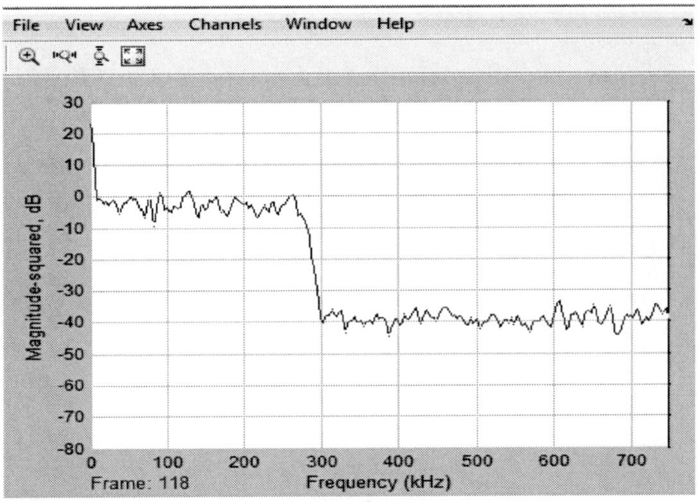

b

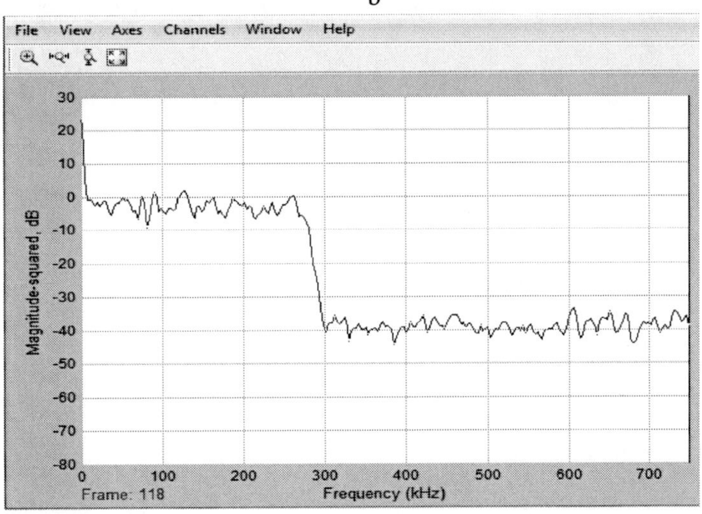

c

Figure 12. (Continued)

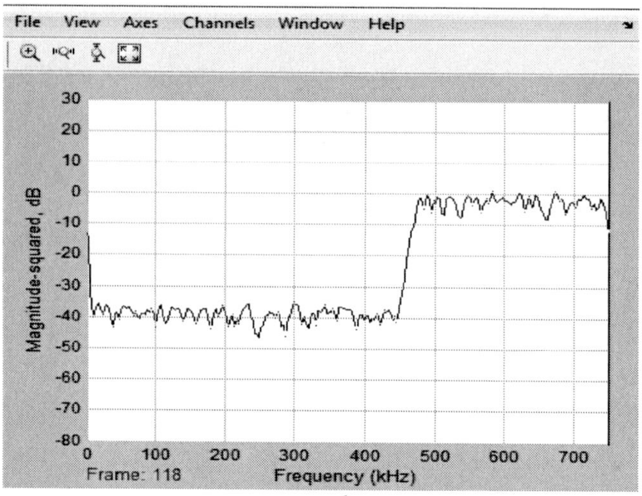

d

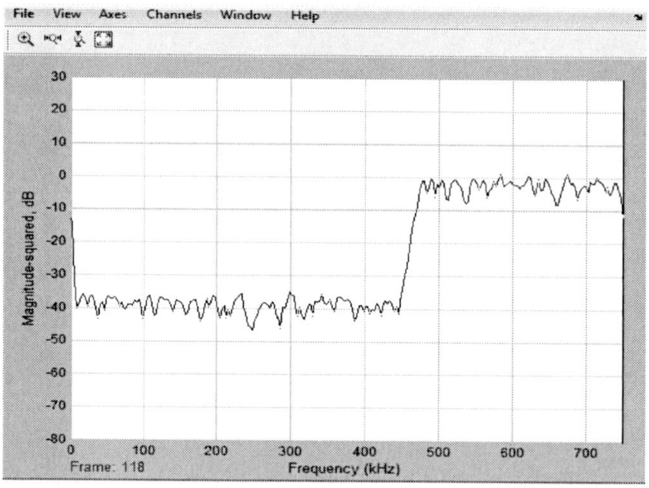

e

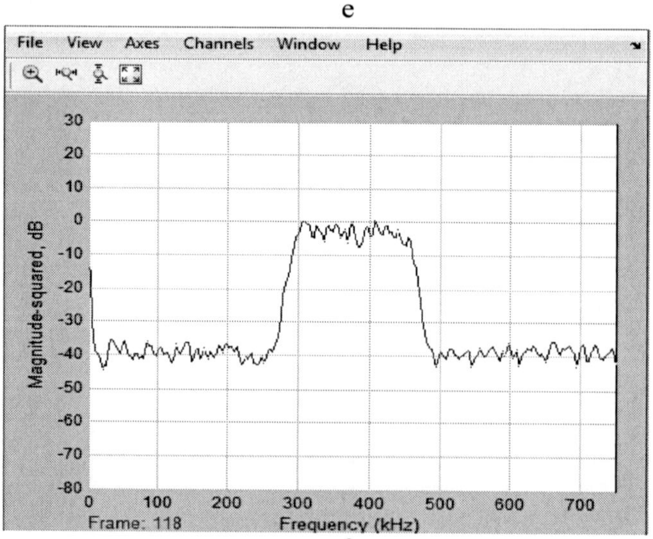

f

Figure 12. (Continued)

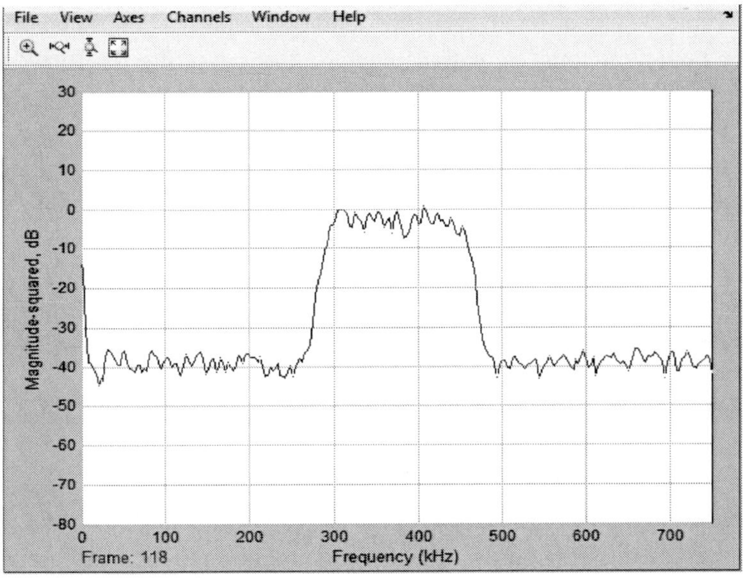

g

Figure 12. a. Input signal using Random source for parallel implementation, b. Output waveform from Low-pass FIR filter using Matlab, c. Output waveform from Low-pass filter using FPGA parallel implementation, d. Output waveform from High-pass FIR filter using Matlab, e. Output waveform from High-pass filter using FPGA parallel implementation, f. Output waveform from Band-pass FIR filter using Matlab, g. Output waveform from Band-pass filter using FPGA parallel implementation.

Table 3. Comparison among Filters

Categories	Low-pass Filter	High-pass Filter	Band-pass Filter	Parallel implementation on FPGA
Minimum period (ns)	10.57	11.16	10.14	**15.03**
Peak memory usage (MB)	294	300	294	**410**
Power consumption (mW)	119	127	118	**273**

DISCUSSION

In this chapter, a ninety two-order low-pass and band-pass and ninety nine-order high-pass FIR filter have been implemented in Spartan-6 LX75T-3FGG676C FPGA board using Xilinx Integrated Software Environment (ISE). The propose idea gives an efficient approach towards the implementation of FIR filter using Simulink and subsequent synthesis on FPGA that suited different applications unlike most previous approaches on a penalty of reducing computation speed. Table 3 shows the comparison among the Filters and their parallel implementation as drafted from the synthesis reports.

From the Table 3 above, cited from the synthesis report of the implementation, it is shown that the minimum implementation time for low-pass filter is 10.57 ns with peak memory usage of 294 MB and power consumption of 119 mW; for high-pass filter, the corresponding time is 11.16 ns with peak memory usage of 300 MB and power consumption of 127 mW; for band-pass filter the corresponding time is 10.14 ns with peak usage memory of 294 MB and power consumption of 118 mW. Therefore, the total period, total memory usage and power consumption for all the three filters when implemented separately are 31.87 ns, 888 MB and 364 mW as against 15.03 ns, 410 MB and 273 mW when parallel implemented on FPGA. Thisanalysis and parameters are used to evaluate the advantages of FPGA and equally validate the effectiveness of parallel implementation of FPGA as earlier discussed.

In addition, below are some of the factors used in evaluating the advantages of FPGA;

1. Time to market - FPGA technology offers flexibility and rapid prototyping capabilities in the face of increased time-to-market concerns. You can test an idea or concept and verify it in hardware without going through the long fabrication process of custom ASIC design.It is easy to implement incremental changes and iterate on an FPGA design within hours instead of weeks.
2. Cost - The nonrecurring engineering (NRE) expense of custom ASIC design far exceeds that of FPGA-based hardware solutions. With FPGA, it means that you have no fabrication costs or long lead times for assembly. This is because system requirements often change over time, the cost of making incremental changes to FPGA designs is negligible when compared to the large expense of re-designing an ASIC.
3. Long-term maintenance - As earlier discussed, FPGA chips are field-upgradable and do not require the time and expense involved with ASIC redesign. Digital communication protocols, for example, have specifications that can change over time, and ASIC-based interfaces may cause maintenance and foreword-compatibility challenges. Due to reconfigurable nature of FPGA chips, it can keep up with future modifications that might be necessary. As a product or system matures, you can make functional enhancements without spending time redesigning hardware or modifying the board layout.
4. Performance - The hardware parallelism nature of FPGAs exceed the computing power of digital signal processors (DSPs) by breaking the paradigm of sequential execution and accomplishing more per clock cycle· Controlling inputs and outputs (I/O) at the hardware level provides faster response times and specialized functionality to closely match application requirements. This again evaluates and validates the parallelism nature of FPGA.
5. Reliability - Due to the fact that software tools provide the programming environment, FPGA circuitry is truly a "hard" implementation of program execution. Processor-based systems often involve several layers of concept to help schedule tasks and share resources among multiple processes. For any given processor core, only one instruction can execute at a time, and processor-based systems are continually at risk of time-critical tasks preempting one another. Because FPGAs do not use OSs, it minimizes reliability concerns with true parallel execution and deterministic hardware dedicated to every task.

Conclusion

This paper discusses the implementation of Low-pass, High-pass and Band-pass FIR Digital Filters on an FPGA. The simulation results show that the output waveforms obtained from the software simulations correspond with those from FPGA implementation respectively using hardware co-simulations. Also, the parallel implementation proves that performance of a low-pass filter is not affected by both high-pass or band-pass filter and vice-versa, and the synthesis report equally shows less resource usage, speed increase, cost effectiveness, more flexibility and low power consumption which validate the parallelism nature of FPGA.

References

[1] Ruan, A. W., Liao, Y. B., Li, J. X., "An ALU-Based Universal Architecture for FIR Filters" *IEEE in proc. International Conference on communications, Circuits and Systems*, pp. 1070-1073, 2009.

[2] P. Panayotatos, "Frequency Response of Filters," © Rutgers University, December 2005.

[3] Abdul Quayyum and Moona Mazher "Design of Programmable, Efficient Finite Impulse Response Filter Based on Distributive Arithmetic Algorithm," *International Journal of Information Technology and Electrical Engineering*, vol. 1, issue 1, Dec. 2012.

[4] Monmasson, E et al. "FPGAs in Industrial Control Applications," *IEEE Transactions on Industrial informatics*, vol. 7, issue 2, pp. 224-243, May 2011.

[5] Wenjing, H., Guoyun, Z., Waiyun, L., "Self-Programmable Multipurpose Digital Filter Design Based On FPGA" In: proc. *IEEE International Conference on Internet Technology and Applications (iTAP)*, PP. 1-5, 2011.

[6] S. Mohanakrishnan, B.E joseph, "Automatic Implementation of FIR Filterson Field Programmable Gate Arrays," *IEEE Signal Processing Letters*, vol. 2 issue 3, March 1995.

[7] Dr. H.J. Manal, H.S. Asaad, "High-Pass Digital Filter Implementation Using FPGA," *IEEE International Journal of Advanced Computer Science and Applications (IJACSA)*, vol. 13 issue 3, June 2013.

[8] "Filter design Methods for FPGAs," Accel Chip, 2006.

[9] Xilinx white paper number 6984, http://www.xilinx.com.

[10] Louis Litwin, FIR and IIR digital filters, *IEEE potentials* 2002.

In: Digital Signal Processing (DSP)
Editor: Juan Zhang
ISBN: 978-1-63485-168-8

Chapter 10

THE INTELLIGENT BAGGAGE SCANNERS

Ivan S. Uroukov*

University of Glasgow, Glasgow, Scotland, UK

ABSTRACT

The increasing importance of obscured object identification is a driving force behind the development of novel and highly technological image analysis approaches.

The process of object identification requires physical recording and analysis of spatial information from the material needing analysis. Here we describe a novel implementation of a spatial image analysis technique that can be used for detection of masked threats such as energetic materials and other potentially hazardous substances which might be carried on board aircraft inside baggage. The image analysis technique is capable of enhancing low contrast obscured objects and identifying their material content. Scene analysis is carried out using textural signatures which correspond to the materials that make up the content of the scene (i.e., baggage). Based on dedicated metrics from a predefined library, the suspicious content i.e., plastic or liquid explosives, or similar, can be recognised. This may trigger imaging enhancement to the acquisition system in the specific region of the luggage.

The detection method can work with a variety of hardware acquisition techniques. However, better results are achieved when a linear scan along an object is employed. This analysis allows system control to be enabled in such a way as to simulate a line scanning geometry in baggage scanners. As a result of the symbiotic relationship between the acquisition method and the 'analyzing core', a range of alarm events could be triggered, allowing further enhanced inspection.

We also demonstrate the application of the technique to other imaging modalities where acquisition is by means of medical ultrasound and security phase contrast imaging for tissue structural and material discrimination purposes respectively.

Keywords: energetic materials, plastic explosives, liquid explosives, weapon detection, material discrimination, baggage scanners, cargo security

* Email: ivan_uroukov@mail.bg , x.mn.901@gmail.com, Ivan.Uroukov@glasgow.ac.uk.

AUTHORS RESEARCH HISTORY

The foundation of the computer vision approach discussed here is an image computing technique which the author began work on in the late nineteen nineties as part of a medical imaging project. It was initially developed for medical ultrasound diagnostic purposes. The image anlysis technique was applied to transrectal ultrasonography for prostate cancer foci detection. The objective was to allow detection of early prostate tumors of a few millimeters diameter. This computational technique was used to evaluate the prostate tissue density and observe the coagulation effects when transurethral microwave therapy was administered during a transrectal ultrasound scan. The imaging method was also successfully applied to detection of mammography structural information. After the initial application to medical imaging, the computational technique was further developed for use in detecting energetic materials, especially plastic and liquid explosives, using airport baggage scanners (Uroukov and Speller 2015). It can also be applied to the detection of drugs and controlled substances such as tobacco. The versatility of the computational approach is illustrated by its application to these various modalities of image acquisition over the period of research reported here.

This approach was further elaborated and combined with an X-ray phase contrast imaging technique (Uroukov et al., 2014*)*. This work has shown the potential of combining technologies for both micro-sized and obscured object detection to overcome some of the limitations of existing baggage and cargo scanning technology.

A further elaboration of the image analysis, refined the computational approaches for use in low contrast electron microscopy (EM) of viruses and large biomolecular. It allowed enhancement and detection of virus structures where other algorithms failed due to the low contrast inherent in cryo-EM imaging.

1. INTRODUCTION

An intelligent baggage imaging system relies on creating X-ray images in an adaptive system based on linear acquisition of baggage as it moves on a conveyer. The image is acquired while conveying the baggage through a scanner device. In general, the scanner comprises of an X-ray source and pixelated detector. The information reading device usually is built of single column, ribbon distributed photo sensors with appropriate X-ray filters and scintillators. More sophisticated pixelisation such as I-ImaS (Griffiths et al., 2008) allows the application of system control for enhanced imaging. Similar approaches (Martin et al., 1999, Huda et al., 2003*)* as I-ImaS are used for medical imaging to ensure minimal patient dose and optimal imaging.

The technique that has been developed in this approach to an intelligent baggage scanning system uses the grey scale image and is based on 'seeing' textures for cues to controlling the system. Textures are more strongly associated with object properties than are the mean grey levels or colour, and hence provide an opportunity to classify materials appropriately. The main difference between the mean grey level measures and the texture analysis is that the texture relies on differences in the spatial arrangement of the grey levels. This provides a technique with additional capacity in comparison to those relying on just mean grey level estimation. The objects in the passenger's bag are more likely to show advanced textural properties (referred to as *textons*) compared to objects comprised of 'pure' materials, where the texture will be more homogeneous. The approach used here has been to consider the bag and its contents to be better characterized with textures. We have used textural analysis to identify suspicious objects and hence adapted the image capture control to maximise the textural information. An emphasis is given to detection of liquids and plastics due to recent developments in threats to the traveling public. Mean grey levels of a baggage scan image are directly associated with the attenuation coefficient of the scanned materials and using dual energy techniques these data can be used to find effective atomic numbers. The textures on the other hand distinguish liquids from non-liquids and one material from another material due to the specificity of their textures. Liquids are usually contained and this implies interference between the grey level due to the liquid and the container. Consequently, imaged liquids in bags tend to appear as textures rather than just homogeneous grey levels. This provides strong support for this approach to searching for liquids and liquid explosives.

1.2. Operational Scenarios

The image analysis (as discussed in research history) was developed independently from the acquisition techniques. For the purpose of baggage scanning, a linear scan (along the luggage) is preferred when conveying the bags through security checkpoints (baggage scanners etc.). A scanning scenario allowing adjustment of the acquisition parameters (such as I-ImaS; Longo et al., 2007) was modelled. It can also reduce the dose by up to 60% in comparison to conventional imaging systems. It is a two-step process (Longo et al., 2007), which uses adaptive analysis of the images to create maximum information within given dose constraints as shown by Griffiths et al., 2008. Firstly a limited area is imaged and based upon analysis of the data the second step involves change or no-change to the imaging parameters for the second 'look' at that area. Making decisions on local regions is central to the operation of such a system and the decision based on textural analysis of the acquired image has shown successful results as reported AL-Hinnawi and Daear, 2012. Here we concentrate on analysis techniques based on textural information to aid in that decision as published by *Uroukov and Speller, 2015*.

1.3. Operational Limits

It is important to relate the object thickness and absorption to its textural content. Indeed, X-ray transmission of the detected signal depends upon the thickness and type of material, i.e., radiation pathlength. However, in most baggage scanning examples the total thickness of the object, i.e., suitcase, does not change much and hence changes in the grey levels relate primarily to the attenuation coefficient of the material.

To address the variation in the content of the luggage and packaging a number of examples were used including industry standards. Example packages were also constructed to demonstrate a general problem. Liquids were also included in the samples.

Generally beam hardening will not be a problem as most baggage scanners operate with a pre-hardened beam quality. Scattering is also minimal due to the use of narrow X-ray fan beams.

2. Background to the Gabor Filters and Their Application to Baggage Scanners

Here we offer a novel approach to characterising materials by their X-ray image textures. We use a Gabor filter bank as a textural 'analyzing core' within an airport security scanning development. In doing this we generate a novel feature space of material descriptors (Uroukov's Code as published by the author in Uroukov and Speller, 2015) based on the textures of imaged material. This space is used to generate a map of targeted materials for feedback to the scanner (I-ImaS type, Griffit et al., 2008) enhanced inspection.

2.1. Seeing Textures

The texture is an important characteristic of X-ray imaged materials. Here we implement textural analysis of materials based upon the spatial variation of the gray-level or mean-signal. The textural features of materials with low atomic number Z, were computed as the 'analyzing core' of a simulated scanning (I-ImaS type) detection system. The research history on textural analysis in imaging fields (Clausi D.A. and Jernigan M. E, 2000) suggests that the human vision system (HVS) may be modeled by means of textural perception shown by Julesz (Julesz B. 1962, Julesz B.1981, Julesz B.2010). The HVS has been found to be sensitive to the frequency, direction and orientation of the image elements. Psychophysical visual experiments by Rao A.R. and Lohse G.L., 1993, showed high-level features for texture discrimination, such as repetition, directionality and complexity, are used by observers. Directionality and repetition are represented by the orientation and frequency, whilst complexity is related to texture consistency.

Furthermore, Marcelia S., (1980) and Daugman J., (1985) modeled receptive fields in the visual cortex of mammals with 2D Gabor functions, yielding indistinguishable residual error with experimental data. Other computational models of periodic and aperiodic pattern selective cells, called grating and bar-cells (found in areas V1 and V2 of monkey's visual cortex) were presented by Petkov et al., (1997*)*. These cells respond strongly to images of bar gratings of a given orientation and frequency but very weakly or not at all to single bars. *Campbell* and Kulikowski (1966) found that humans have orientation and spatial frequency sensitivity, which can be modeled as a number of independent detector mechanisms; each represented by a narrow band filter tuned to a different frequency. Gabor filters have both frequency and orientation selective properties. They possess optimal joint resolution in the spatial and frequency domain (Daugman J., 1985 and Jain A.K., 1991) and are therefore suited for modeling of the HVS.

2.2. Textures and Gabor Filters

Research has demonstrated that the human vision system generates a multi-resolutional decomposition *(Marcelja 1980, Daugman 1985). Sin*ce wavelets are intrinsically multi-resolutional, they have been implemented successfully in texture analysis models (*Jain, 1991)*. A popular multi-channel filter is the Gabor function (*Gabor, 1945*). It is not truly a wavelet but may be implemented in such a manner as to mimic the properties of a wavelet and hence provide a good mimic of the behavior of simple cells in human vision. It has appealing simplicity and optimum joint spatial to spatial-frequency localization. The filter has the ability to isolate specific frequencies and orientations, which makes it suitable for texture analysis. In terms of functionality, a Gabor function is a Gaussian modulated sinusoid. It is defined as the product of an elliptical Gaussian with aspect ratio: σ_x / σ_y, with centre:

(X, Y)) and a complex exponential.
Where:

$F = \sqrt{U^2 + V^2}$ is spatial frequency,
$\theta = \tan^{-1}(V/U)$ is orientation.

The complete Gabor function can be represented spatially by:

$$f(x,y) = \frac{1}{2\pi\sigma_x\sigma_y} e^{\left(-\frac{1}{2}\left(\frac{(x-X)^2}{\sigma^2{}_x} + \frac{(y-Y)^2}{\sigma^2{}_y}\right)\right)} e^{(2\pi j[U(x-X)+V(y-Y)])} \quad (1)$$

When Gabor functions are applied as linear filters, the centre defined by (X, Y) in the spatial domain is unnecessary, since the function will be implemented using convolution. In the spatial-frequency domain the filter looks like a Gaussian centered at (F, θ), hence Gabor filters are band pass filters. A complete (even and odd components) Gabor filter can be represented spatially as follow:

$$h(x,y) = \frac{1}{2\pi\sigma_x\sigma_y} e^{\left(-\frac{1}{2}\left(\frac{x^2}{\sigma_x{}^2} + \frac{y^2}{\sigma_y{}^2}\right)\right)} e^{(2\pi Fx)} \quad (2)$$

In texture research (see works of *Jain and Farroknia, 1991*) it is widely accepted that a bank of real (even-symmetric) components of a Gabor filter can be used to characterize the channels. The justification of using an even-symmetric filter is on psychophysical grounds but little evidence is found to support this basis. The representation is as follow:

$$h(x,y) = \frac{1}{2\pi\sigma_x\sigma_y} e^{\left(-\frac{1}{2}\left(\frac{x^2}{\sigma_x{}^2} + \frac{y^2}{\sigma_y{}^2}\right)\right)} \cos(2\pi Fx) \quad (3)$$

Filter Implementation

For our purposes, the filter has been built as ‘cos’ and ‘sine’ masks and the complete mask is obtained by point wise multiplication. The masks are *Cos.Mask* and *Sine.Mask*:

$$Cos.Mask(x,y) = C_N e^{\left(\frac{-r^2}{2\sigma^2}\right)} \cos(\varpi x') \tag{4}$$

$$Sin.Mask(x,y) = C_N e^{\left(\frac{-r^2}{2\sigma^2}\right)} \sin(\varpi x') \tag{5}$$

where,

$x' = x\cos(\theta) + y\sin(\theta)$, $r^2 = x^2 + y^2$, $\varpi = \frac{2\pi}{T} = 2\pi f$, C_N - normalization constant

Complete gabor mask is defined: $Gabor.Mask = Sin.Mask * Cos.Mask$.

As an example, consider the case of a circular Gaussian $\sigma_x = \sigma_y = \sigma = 6$ [pixels]. The size of the Mask (*Mask.Size*) is equal to 51 pixels, i.e., 51x51 pixels and the period $T = 6\sqrt{2}$. The radial frequency is $\varpi = 2\pi / 6\sqrt{2} = 0.74[revol./width - img]$ and orientations have been chosen for this example as θ = 18°. When the psychophysical approach for the frequency span is to be used, it is recommended that the ratio T/σ be set to $2\sqrt{2}$. A representation of the mask where the Gaussian is circular is shown in Figure 1.

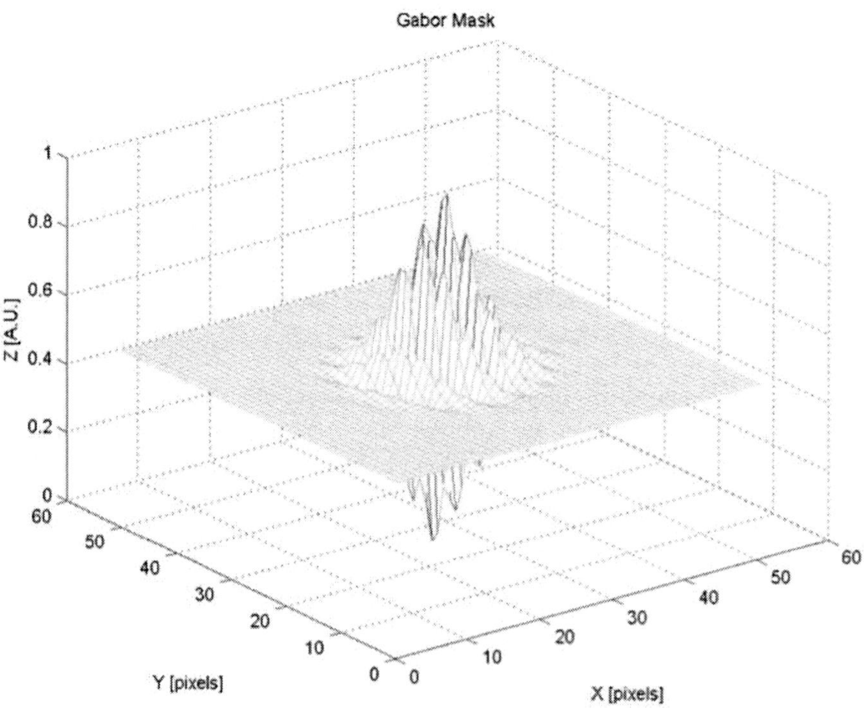

Figure 1. Masks of Gabor filters for a given set of operating parameters.

Filtering is a convolution of the image with the mask, where the parameters $\varpi,.\theta$ are set in advance. A decision on the circularity of the Gaussian has been influenced by fingerprint analysis where circular Gaussians were used (Lin Hong et al., 1998) along with the selection of a value for σ (the standard deviation) which involves a trade-off for a circular Gaussian, where $\sigma = \sigma_x = \sigma_y$.

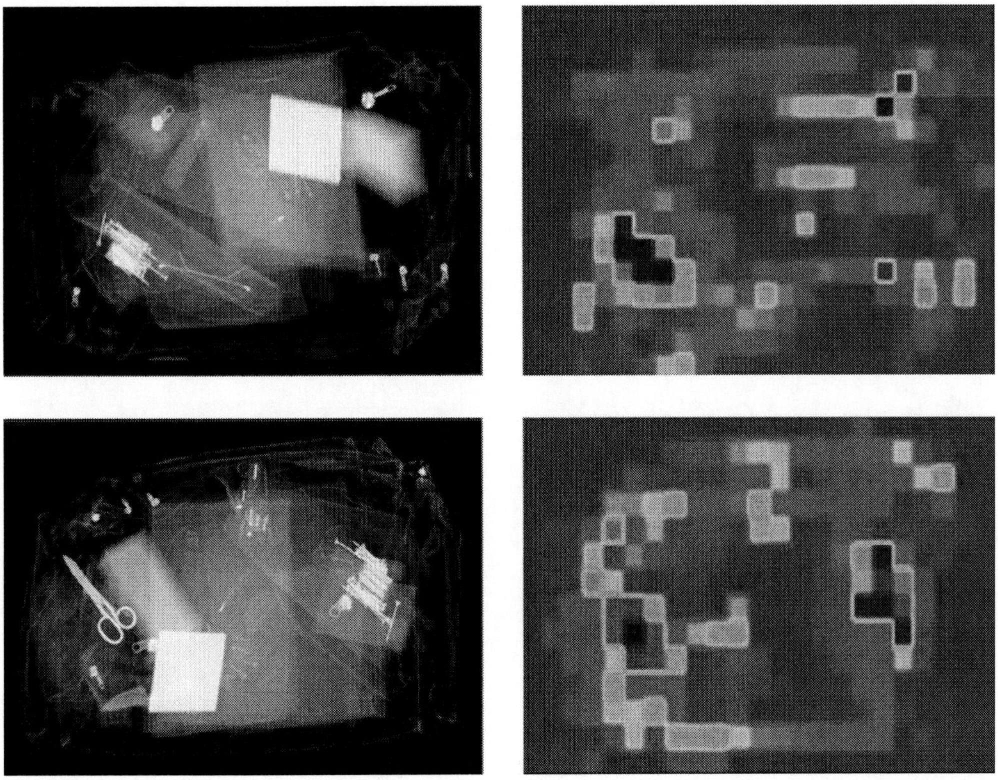

Figure 2. Directional field computations applied to images taken from a 3DXray Ltd baggage scanner operated in monoenergetic mode. Note the ability of the filter to identify two suspicious regions in the images.

2.3. (B3) Directional Field Computation

Direction Field (*DF*) is a good descriptor of pointy and sharp objects, which are good descriptors of many weapons or weapon components. The objects tend to express directed textures in the image; therefore this fact can be used in their estimation. The DF is, in principle, perpendicular to the gradients in the image. However, the gradients are orientations at the pixel scale, while the DF describes the orientation of the structures depicted on the image in a much coarser scale. Therefore, the DF can be derived from the gradients by performing some *averaging* operation on the gradients, involving pixels in some neighborhood (in a window). The gradient-based method was introduced by Kass et al., 1987 and has been adopted by many researchers. To evaluate how well they point in one direction, we compute the coherence.

$$Coh = \frac{\left|\sum_{W}\left(G_{s,x}, G_{s,y}\right)\right|}{\sum_{W}\left|\left(G_{s,x}, G_{s,y}\right)\right|} = \frac{\sqrt{\left(G_{xx} + G_{yy}\right)^2 + 4G_{xy}^2}}{G_{xx} + G_{yy}} \tag{6}$$

where, the squared gradients are in a local window W are given as follow:

$$G_{xx} = \sum_{W}\left(G_x^2\right),\ G_{yy} = \sum_{W}\left(G_y^2\right),\ G_{xy} = \sum_{W}\left(G_x G_y\right), \tag{7}$$

$(G_x G_y)$ is defined as $\left[\frac{\partial I(x,y)}{\partial x}, \frac{\partial I(x,y)}{\partial y}\right]^T = \begin{bmatrix} G_x(x,y) \\ G_y(x,y) \end{bmatrix}$ which may be used to estimate the direction and defined as :

$$\begin{bmatrix} G_x(x,y) \\ G_y(x,y) \end{bmatrix} = sign\left(G_x(x)\right)\nabla I(x,y) = sign\left(\frac{\partial I(x,y)}{\partial x}\right)\begin{bmatrix} \frac{\partial I(x,y)}{\partial x} \\ \frac{\partial I(x,y)}{\partial y} \end{bmatrix} \tag{8}$$

The local gradients of a gray-scale image *I (x, y)* are calculated by means of a convolution with the derivative of a Gaussian window. An example of this technique applied to a baggage scan is shown in Figure 2.

3. Understanding Textons. Model of Visibility. *Uroukov's Code* and X-Ray Vision

Two fundamental parameters have been considered in developing these techniques for controlling the scanner – frequency and direction. The combination of these parameters and their subsequent *textons* has been used to identify suspicious objects and to adapt control of the recording system through optimizing visibility. Experimental determination of textons for all materials imaged under a wide range of conditions is beyond the scope of any pilot study. Instead we chose to investigate the composition of textons for particular materials (organics, plastics, liquids) which are closely related to those we are interested in (i.e., plastic explosives etc.). Although a full study of all these materials could not be undertaken due to time constraints, an approach has been developed to study the problem. The approach uses a statistical estimation of a single parameter to represent the output of the filter bank after operating over a composite image comprising of multiple interfering textures produced by materials of interest and their containers.

The recorded images have been segmented according to the image contents and the subsequent Regions of Interest (ROI) were analyzed before and after filtering. Segmentation itself may be conducted in manual or automated manner. The gray levels within these ROIs are averaged (hence Mean Gray level per ROI (MGR)) for no filter and for each filter bank parameter. MGR's have been evaluated for the two filter parameters i.e., orientation and frequency and stored in a 3-D space (see below). Figure 3 shows how the MGRs are obtained from a typical image.

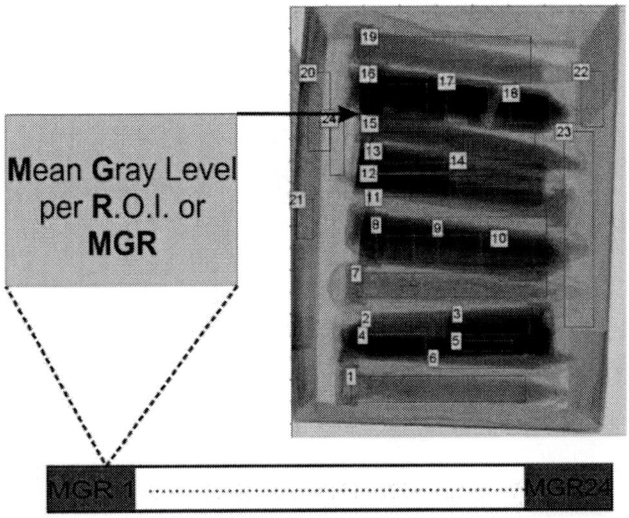

Figure 3. Conversion of ROIs into MGR. This has been carried out for different values of filter orientation and filter frequencies.

The MGR values are organised into a 3D space (Uroukov's Code) as shown in Figure 4.

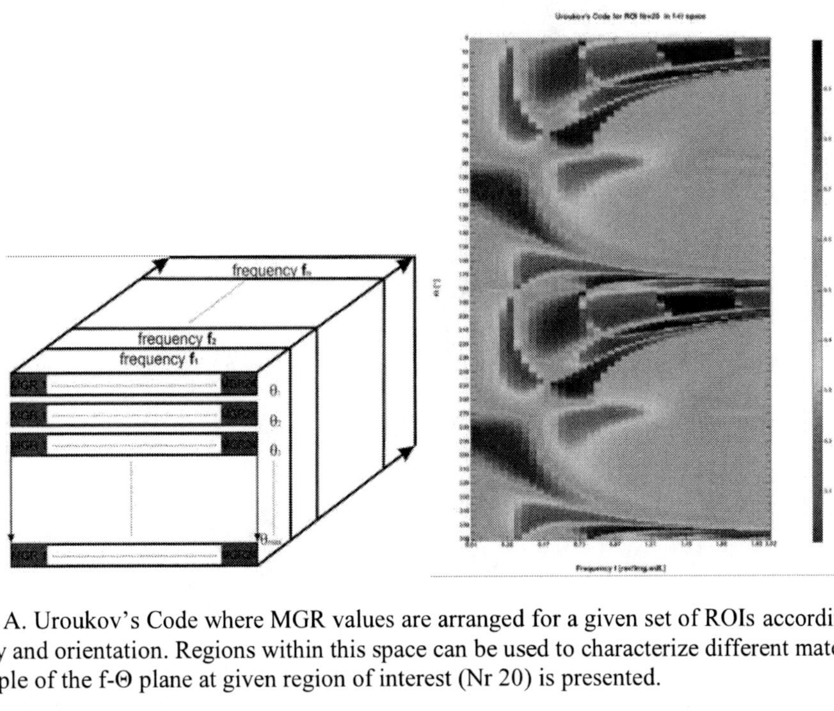

Figure 4. A. Uroukov's Code where MGR values are arranged for a given set of ROIs according to frequency and orientation. Regions within this space can be used to characterize different materials. B. An example of the f-Θ plane at given region of interest (Nr 20) is presented.

When the MGRs are analyzed, it appears that organic substances are localized in very similar locations and they may be taken as an ensemble that possesses similar characteristics as shown in Figure 5.

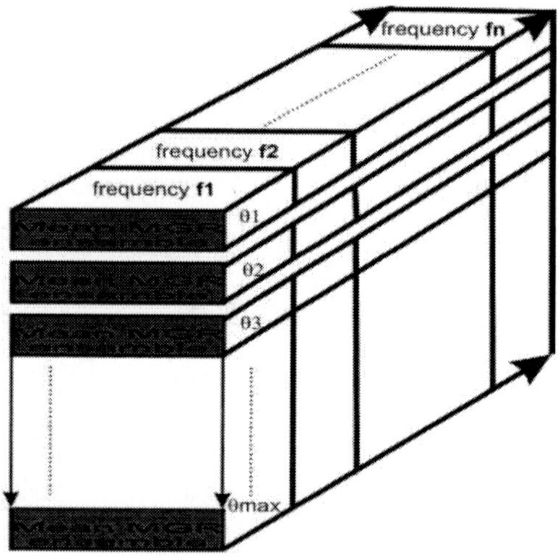

Figure 5. Condensation of Urokov's Code into regions of similar materials.

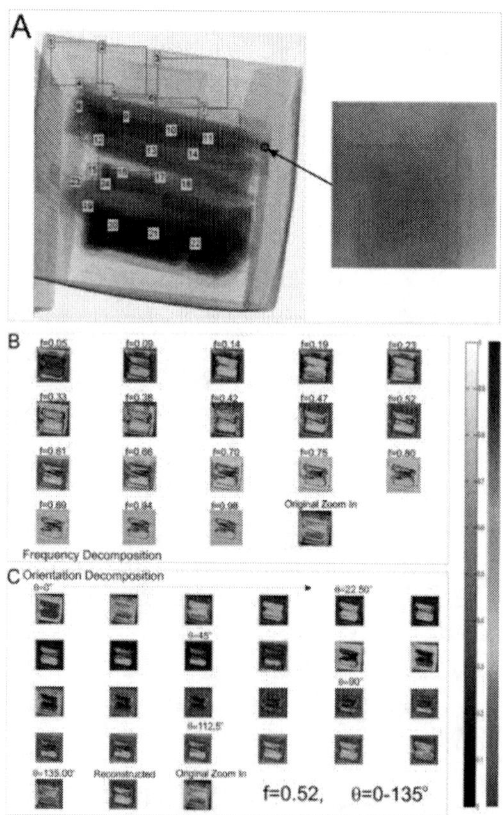

Figure 6. Images of the test object called 'Boxed Pack' and magnified area demonstrating a textural pattern. An ROI has been selected and filters of different orientation and frequency been applied. Note the ability of the filters to select materials.

To understand the behavior of organic substances, Figure 6 shows the results for a test image- box and containers filled with various organic substances. Note the organization of the ROI taken and their signal levels in respect to the background and the rest of the sample for restricted ranges of orientation and frequency, i.e., theta = 70-90° and f = 0.5-08.

The MGRs obtained from Figure 6 are plotted in Figure 7 for the sum of all orientations per frequency per particular ROI and in Figure 8 as the sum of all frequencies per orientation per ROI.

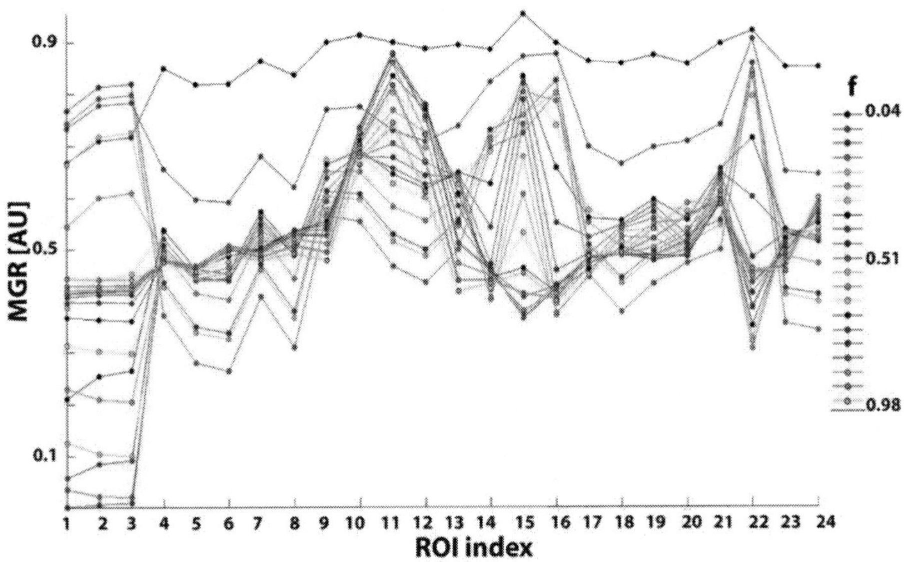

Figure 7. Plot of MGR summed over all orientations of the filter for a particular frequency as a function of ROI.

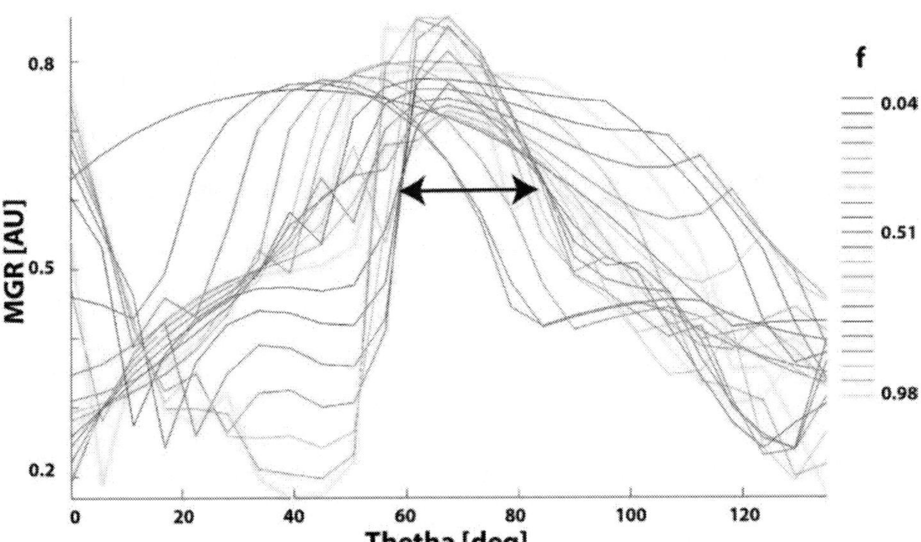

Figure 8. MGR values for a range of different orientations of the filter. Note the clear maximum in MGR for a range of orientations.

In Figure 7 it is clear that at ROIs 1-3 (packaging only) levels vary significantly per frequency but for ROI's 4-18 they maintain values of 0.6-0.8 for all frequencies. Similar behavior is observed for the second group of ROI19-22. Thus, particular frequencies are selective for particular ROI's.

Figure 8 shows a similar analysis but for orientation of the filter. Optimum orientations are indicated as within the range of 55° - 85°. To demonstrate this, the same image of organic substances is shown in Figure 9 but at selected orientations and frequencies where the contrast between the targeted substance and background is 100%; i.e., the filter has selected the particular substance, which in this case is aligned between 60 and 90 degrees.

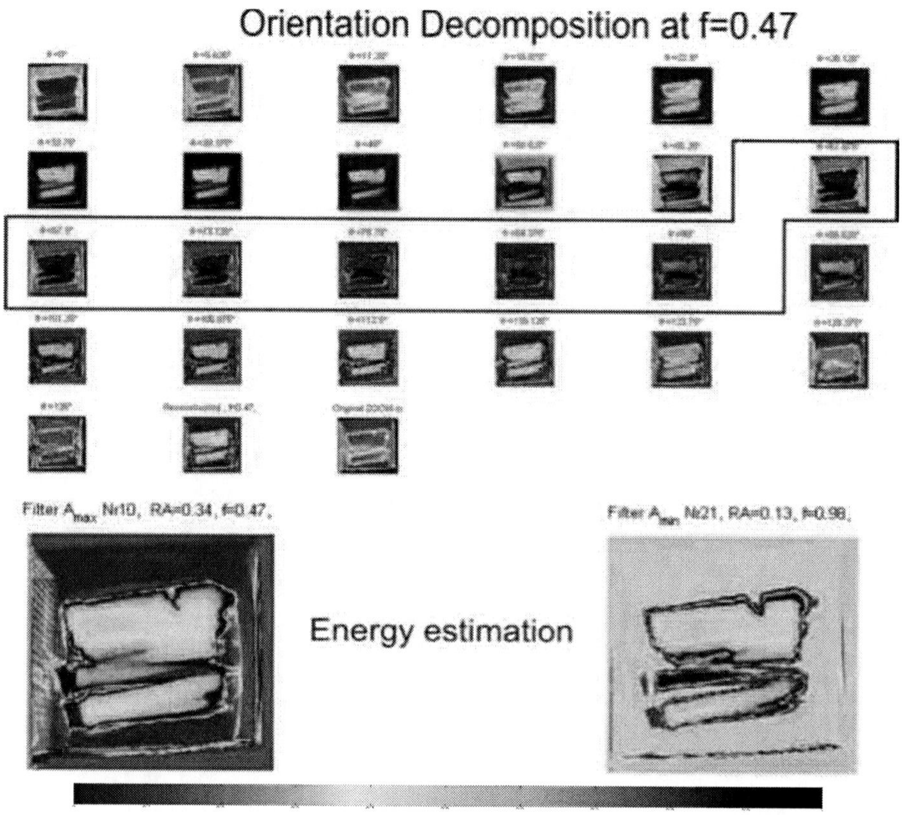

Figure 9. Effect of selected orientations and frequency for material selection. Note the local minima when the MGR values are plotted. Color bar refers to image color map in the scale of intensity 1-100%.

The use of the optimum frequency range produces the result shown in Figure 10. If an approach that reflects the particular substance is taken, then the descriptors of the resultant textures as integrated levels can be used. Here, in Figure 11, we demonstrate two of those descriptors, which are obtained for each ROI and show MGR distribution v.s. orientations per frequency. It is assumed that those descriptors describe textures coming from the ROI, i.e., it is not only describing one material as the ROI contains other textures (albeit in small quantities) from the packaging and container, etc. As can be seen in Figure 11, ROI 20 and 21 have very similar descriptors but ROI 14 and 16 are quite different and hence contain properties that are characteristic.

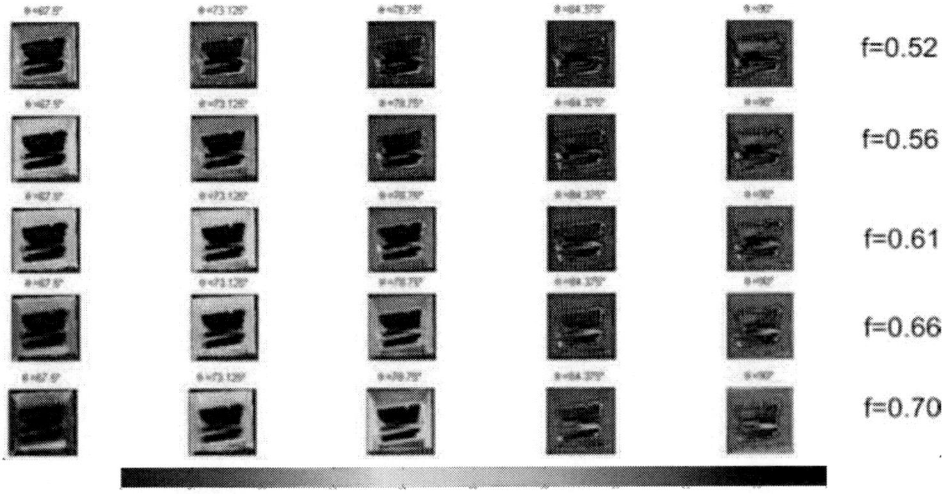

Figure 10. Use of the optimum frequency range to select materials. In this case organics are selected.

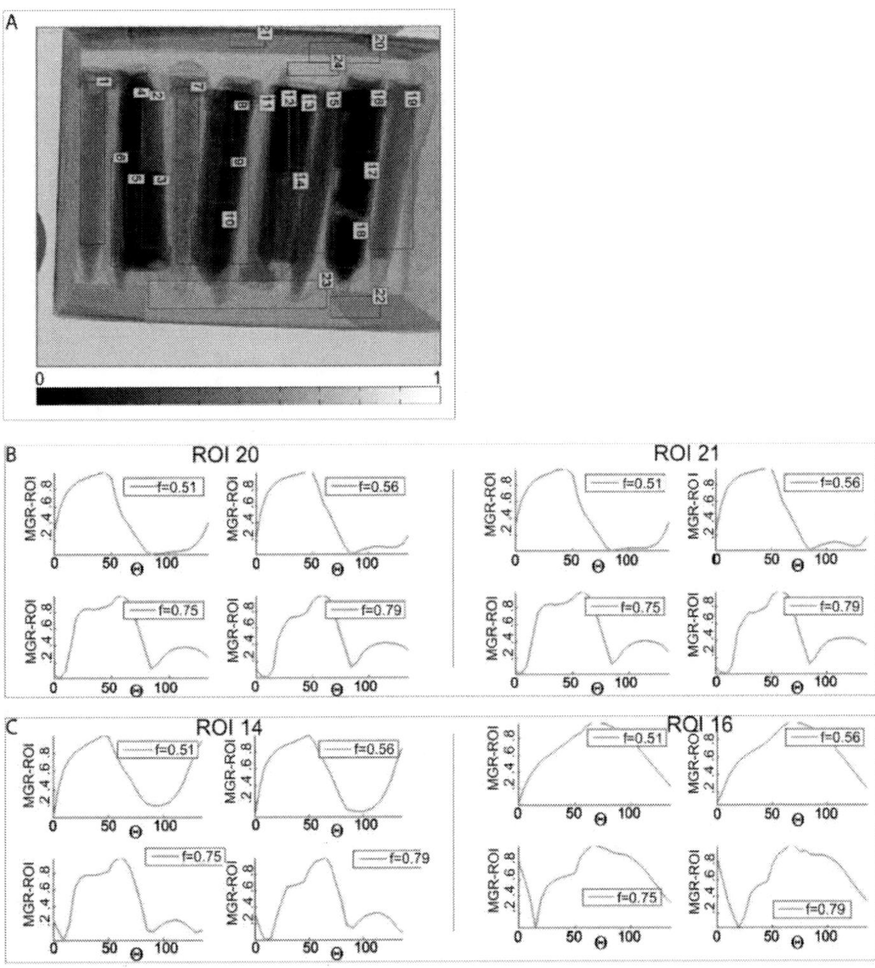

Figure 11. The image and descriptors for different ROIs within the image. Note the similar behavior of some ROIs leading to the characterization of those materials.

3.1. 'YAUW' Method for Detection of Substances

This method was developed to detect virus structures based on their textural signatures. Viruses are comprised of various biomolecules. Those are built from very similar chemical substances, such as nitrogen, oxygen, carbon, and phosphorus. It was shown that biomolecules of interest (Hemagglutinin and Neuraminidase) possess some characteristics in terms of frequency and orientation ranges of the analyzing core (filterbank) in the extension of the algorithm discussed here. When a *'Uroukov's Code'* is derived, the algorithm may look for localizing extremes in this parametric space. Those extremes are specific, and correspond to a specific range of substances and structures associated with their textural properties in the image. If suspected ROIs with similar substances are sampled (selected as ROIs), and integrated then we will obtain an ensemble of similar substances. An integrated MGR is derived from this ensemble as a function of orientation per frequency.

The following step is to derive the gradients and locate the rapid slopes in the integrated MGRs. As a result of this, a detection function is generated to indicate the located gradients to the orientation range. This algorithm is used to generate a range of orientations per frequency for the selected ROIs. These orientations correspond to specific features, which are obtained by secondary reconstruction of the image. If an ensemble of frequencies is used then the range may be extended depending on the predefined precision. Figure 12 illustrates the computational process on a sample with a frequency index of 20.

Dedicated ROIs were integrated for MGR response at f_{20}. The result is presented on Figure 12 (upper graph -1/3, label - MGR at f_{20}, blue line). Then a derivative was calculated and slope spots were localized as shown by the red line (label – Derivative of MGR). A detection function is generated in response to the derivative shown with green line (label – Detection Function). The total slope detection function presents the 'hot' orientation spots where reconstruction may be triggered further.

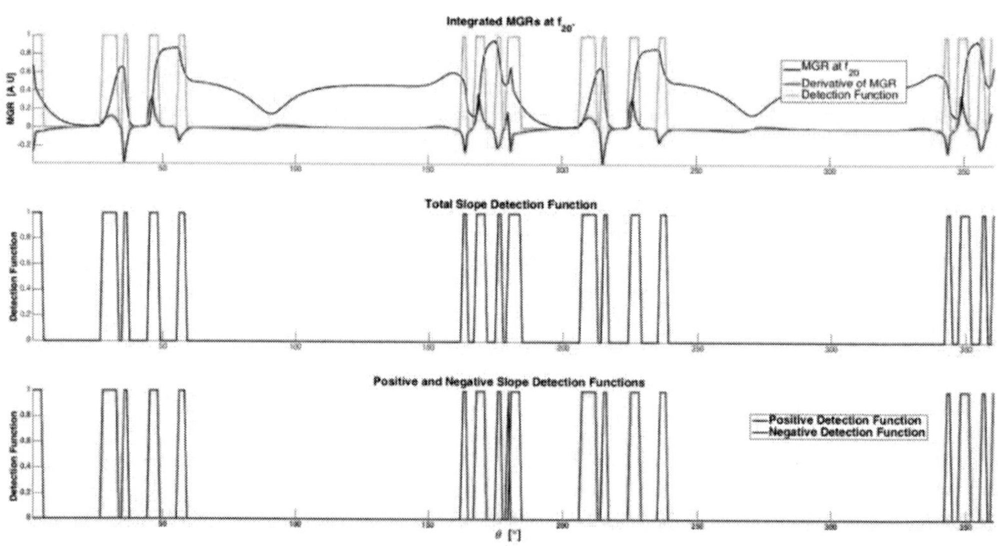

Figure 12. Illustration of 'YAUW' Detection Method.

These are shown in Figure 12 (middle graph -1/2, blue line). The author has found interesting results for positive and negative detection functions. Reconstructions were performed for both gradients. However, a combination of both – positive and negative gradients was also considered although the total detection function was used for orientation range generation. The positive and negative detection functions are shown in red and blue lines respectively on Figure 12 (lower graph -1/3, label Positive Detection Function, Negative Detection Function). This approach was applied in virus analysis for detection of biomolecules such Hemagglutinin and Neuraminidase. This computational approach can be implemented in a similar manner for more precise material discrimination using various scanning techniques.

4. Model of Visibility

The system we propose will operate by imaging an object, finding suspicious regions and then adapting the system to maximise visibility of the suspicious region. Hence the purpose of the system is to enhance detection and thus it is important to define the optimal visibility where the detection algorithms will operate. Visibility is a term that may be defined in many areas of neurology and ophthalmology as an optimal condition of perception of visual information. In optical and neuropsychological fields, to obtain the visual information and recognize the portion of it that is the subject of interest amongst other visual information, is defined as visual ability. In the field of computer vision, visual ability is less well understood and not well defined. Perception of the visual world uses some of the most complex processes associated with human brain function. It involves multiple processing of the visual scene through many stages before the information arrives at the visual cortex in the form of 'trains of ionic spikes or currents' that may be recorded with intracellular electrodes.

It is well known that such multistage processing is performed in the retina by a complex of cells organised in multi-layers and interconnected. They are localised behind the photosensitive rods and cones, which acquire intensity and colour related spatial information per subarea (pixel like). It is well known that information may travel omni-directionally along the *nervus opticus*. The multilayer processor at the eye site receives its control from the brain and modulates the transmitted information before it travels to the visual cortex. A model demonstrating the inhibitory (red) and excitatory (blue) pattern across the retina is presented here in Figure 13 ('Eye Design'). It is based on Gabor-like kernel configurations, which are frequency and orientation dependent. Furthermore, as in high-level cognitive functions, modelling is involved in terms of simplified mathematical processes. However, this is a simplification of the physiological processes, as the exact pathways involving reception and neuro-computing are not well understood.

Due to these simplifications, the study of vision and perception is approached without invoking precise computing in the brain, but as a system that has an input, an output and a simple model of electrophysiological behaviour associated with a particular study. Perception and psychophysics use models of perception that involve elongated fields in the V1 and V2 areas of the visual cortex. These may be either excited or inhibited when stimulated by observation of elongated textons.

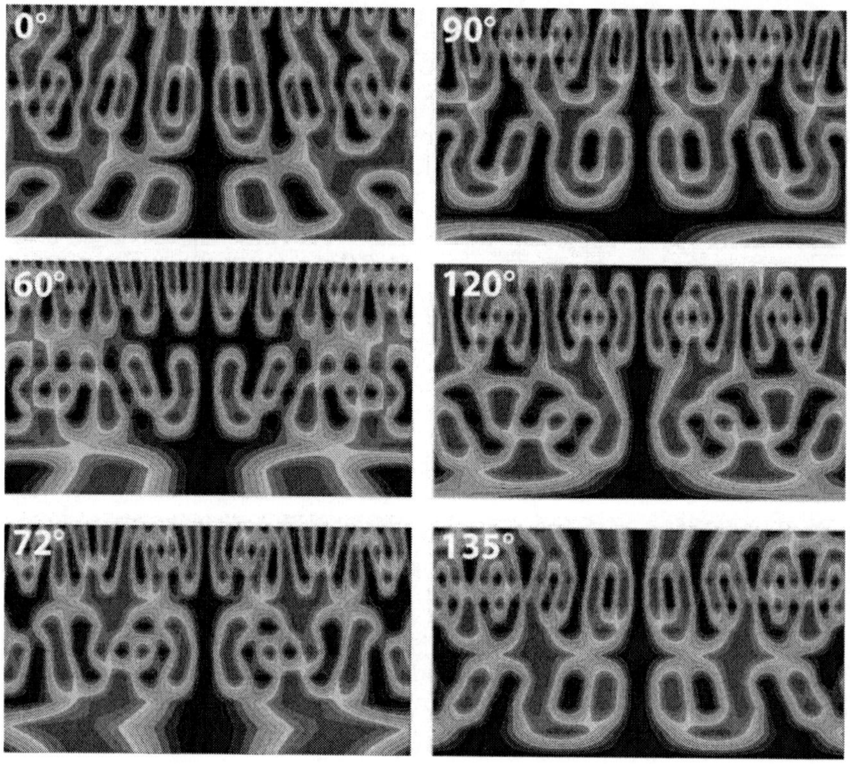

Figure 13. 'Eye Design'. Sequence of Gabor-like kernel configurations shown at 0°,60°,72°,90°,120° and 135° orientation parameter.

These regions have been found to respond strongly to textures comprised of elongated and oriented textons. Therefore, our scanning system has applied a model that is physiologically inspired and mimics one of the perceptual mechanisms found in mammals.

There are a number of classes of mathematical function that are associated with similar patterns of blobs or spots i.e., textons, and the convolution of these functions with the pattern of the textons in the image is one of the perceptual mechanism implemented in computer vision research.

5. Uroukov's Code and X-ray Vision

As discussed Uroukov's Code (a reconstructed volume bounded by frequency and orientation) is a model that is inspired by perceptual principles in the real-word. Within this space there are two basic directions. We can decrease the degrees of freedom in understanding real-world vision tasks to these two directions in order to implement simplified operational procedures on a computer.

The first approach: 'A' -Principles from *Real-life World* leading to principles in the *Machine World* leading to the *Mathematical/Abstract World* or spaces in which we compute the visibility measure.

Second approach: 'B' - The associated processes in the 3 spaces provides the highest level of dimensionality as in the real-life world. The processes in their activity spaces, are limited by particular space dimensions, however they are associated in all three spaces although they are simplified in the numerical definition as it leads to a single number or threshold.

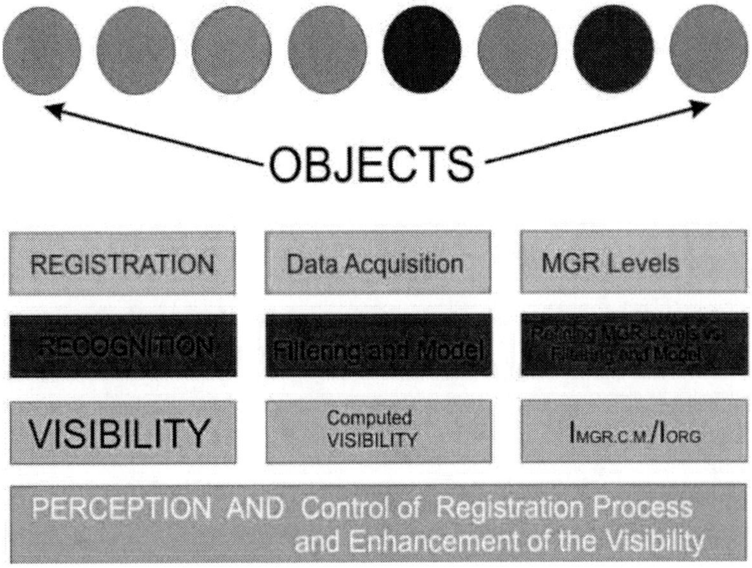

Figure 14. Description of a visual problem as three spaces.

Figure 14 demonstrates the three spaces and therefore contains the maximum dimensionality.

For approach **A** this becomes:

- The first is '*Registration-Recognition-Visibility'* in the real-life world as it includes both speed and the highest level of brain cognition
- The second is the one that applies to the machine world and is '*Data Acquisition – Filtering and Model – Computed Visibility Vector'*
- And the final, numerical world is associated with 'MGR Levels – Refining the MGR Levels vs. the Filtering and Model and $I_{MGR.C.M.}$/ I_{ORG} '

For each of these stages we have similar processes which are:

- The initial registration of the object that is associated with a particular signal level and leads to a mean gray level for a given ROI.
- The recognition process is high level information processing and is the result of multiple neuronal network computing before the information reaches the visual cortex. In our model this process is represented by the use of the filter bank.
- The recognition/characterisation of objects is the final outcome before perception. In our case it is the signal that allows quantification of the process and is used at defined thresholds to perform a perceptual function. We use the normalised computed visibility, and selectivity is based on the textural content that is above a given threshold of recognition.

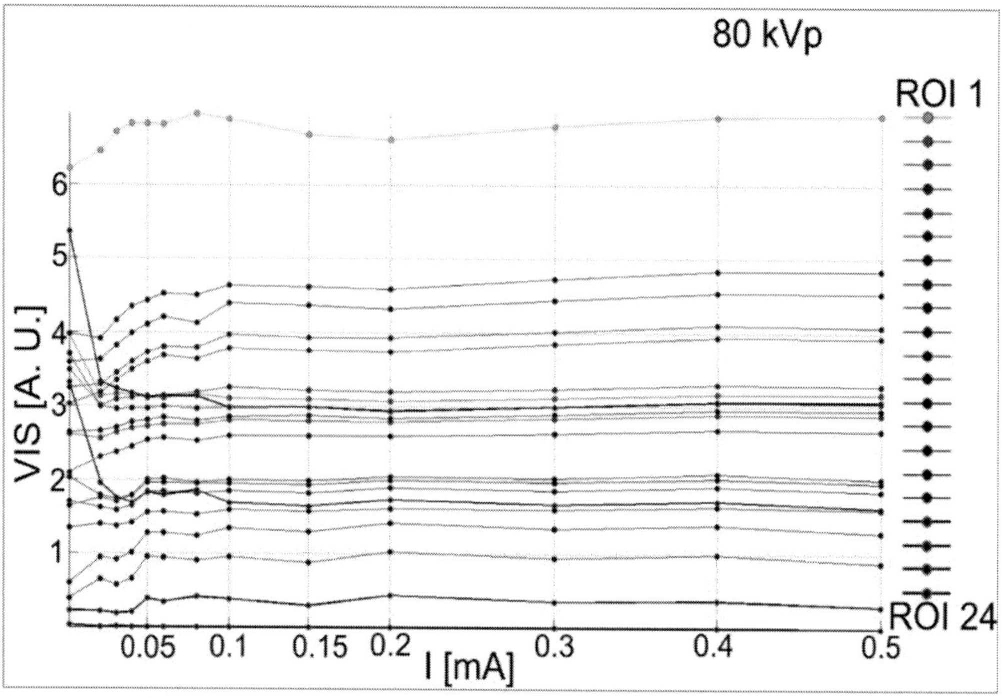

Figure 15. Illustration of computed visibility per ROI or material for different X-ray tube currents.

Figure 15 demonstrates how visibility is affected by the adaptable controls on the X-ray imaging system. In this case the tube current was varied to alter the noise content of the image. It can be seen that the visibility is not consistent for all the ROIs and depends on the filter bank selectivity. This is associated with perceptibility and depends on the targeted object. A high cognitive element is involved before the visual process is accepted for recognition in the brain. It is well known that in the eye, following detection by receptors in the retina, the information, as action potentials, is gathered in layers of neurons behind the photosensitive layers. These may have many inputs but only a single output leading to the optic nerve.

Some physiologists explain this as *performing a spatial integration of the received image in the eye*. It should be understood that this processing may involve excitation/inhibition and may stop in time or may modify the spatial information before it enters the *nervus opticus*. In some cases physiological evidence has shown that for real perception and high, multiple levels of reception, the visual information seen by the eye may not be transmitted to the brain. This process is still under study and may be an area of future study in machine vision. In the baggage scanner the targeted object or 'subject of interest' may involve processing similar to that involved in the human eye but it is a matter for additional research whether this inhibitory/excitatory mechanism needs to be involved in the design of feedback control. This will increase the selectivity of the imaging system. The major difference between the feedback used on our scanner and the one originally employed in I-ImaS is the mechanism on which the feedback operates. We have attempted to involve more complex and 'high cognitive like' control cues to achieve high selectivity in the search for explosives during airport X-ray inspections.

6. Simulation of the Scanning System

The experimental imaging system was based on the tungsten target source X-Tek, SR125. However, several limitations to this approach were found when using such an X-ray source, which has limited power.

The source to detector array distance allowed a magnification of 1.05x, which was used for all test objects. The arrays used were 40 x 30 cm dual large area detectors (Pax Scan supplied by Varian). These detectors could simulate all configurations of baggage scanner sensors, and allowed scout/real beams to be simulated. Before any images were taken the X-ray tube was conditioned and the Pax Scan system allowed to stabilise. The imaging array was calibrated for offset and gain of the sensors, and for defective pixels, using an X-ray tube current of 100 μA. This process was carried out for a range of anode voltages from 35 kV to110kV. Once calibrated, flat field checks were carried out at regular intervals to ensure that no change had occurred in tube field uniformity or array performance.

6.1. Imaging Procedure

All test objects were imaged under a range of conditions that might be created by the adaptive control of the scanning system. Thus, at each anode voltage, a calibration at 100 μA was carried out, then a set of images was generated using anode currents between 10 μA, and 300 μA, or in some cases between 50 μA and 500 μA. For each image, an ROI was selected for recording the background noise. With the wide range of anode currents (some well away from the calibration values used) a degree of noise is introduced into the images. It is important to note that detection algorithms are not sensitive to the noise, which is introduced to the imaging system by tube current variation. As a further check on imaging system stability an ion chamber was used to record detector exposure for all images.

6.2. Image Resolution

For an experimental validation, the image resolution (sensor Pax Scan 4030A) of 2300 x 3200 pixels was kept. The achieved pitch size was 6.7 times smaller in comparison to standard airport scanners. This guarantees the reliability and operation of this development on scanners with lower resolution. It turned out that detection was unaffected over this range and the simulations were reported with 800μm pitch size, which is the most frequently used sensor pitch in current baggage scanners.

The converted and uncompressed TIF image was digitised with a resolution of 2^{16} levels of grey. This resolution is computationally and optically good, although the human visual system does not distinguish more than 64 or 2^5 levels of grey.

7. Relation Between Filter Detection and Variations in the X-ray Tube Operation

A fundamental question is whether or not the filter performance depends upon variations in imaging factors caused by hardware instability i.e., fluctuating of the X-ray source output due either to tube current or tube potential changes. To understand this we carried out full sensor calibration at only one tube current (100μA), although many images were taken at different tube loadings. In this way, the detector performance is slightly shifted from the compensated mode and a level of noise was introduced. Two checks were carried out, one where tube currents were changed and the other where tube potential was altered.

Images of the test sample were produced at 60 kVp. The current was changed between 10 μA and 300 μA. The detection levels were satisfactory to allow imaging of the minimal amount of textural information and the filter bank demonstrated stable operation.

As the result shows the signal left after filtering, it is concluded that any filtered images obtained with at least 80% of the inherent signal guarantee stability for the filter (Figure 16). A measure of this efficiency is obtained by observing how the standard deviation (values in red triangles) changes compared to changes in the magnitude of MGRs (values in black circles). The standard deviation (STD) is significantly lower for experiments conducted with 100 μA and 300 μA. As the STD decreases it signifies the deterministic character of the MGR peaks and therefore it signifies the constituted output of the filter bank. This is a guarantee of the measure of the selectivity of the filter bank. As we observe this behaviour with increasing precision above 100 μA, we conclude that all image features are well detected but the signal approaches its maximum when a tube current of 60 μA or more is used.

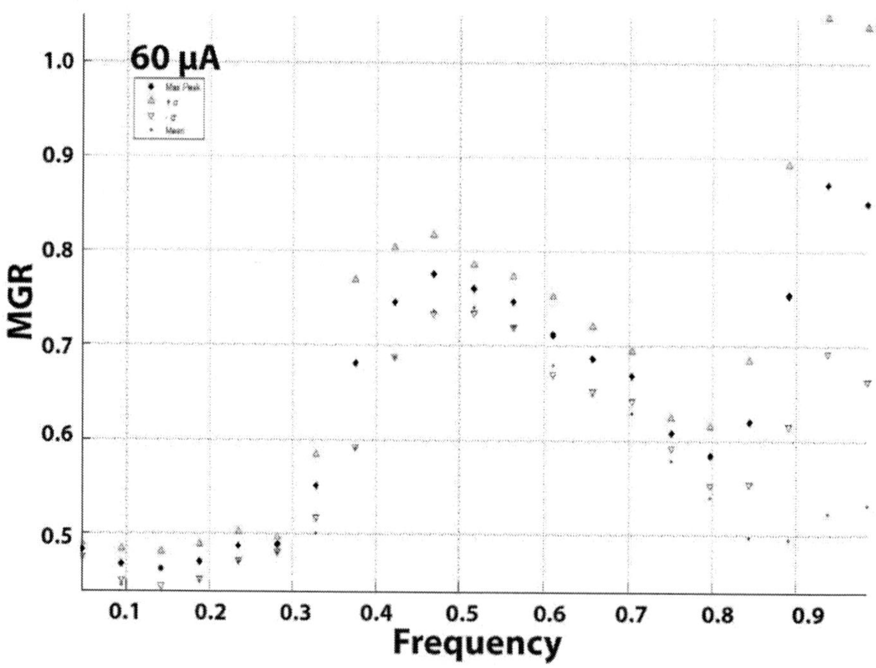

Figure 16. Texture evaluation at 60 μA imaging.

8. System Performance

To evaluate the performance of baggage scanning, the system was tested against its ability to improve the detection of organic materials, which are often mis-classified in current dual energy baggage scanning units.

8.1. Image Decomposition

The first part of operating a baggage scanner system would be decomposing/segmenting the image into suspicious and normal regions. It was shown that orientation and frequency decomposition methods yield useful information for material characterisation. Artificially designed and created samples were used. Briefly, a number of tubes with organic contents were packed in a box and imaged in different directions, as might be the case in a real scanning situation. Four independent image sets were created - the front, the back, the left hand, and the right hand sides. Analysis showed an efficient performance of the decomposition process.

In carrying out material characterisation/selection there are a number of processing stages. These stages are described below and each stage starts a new page in the description. Two decompositions were performed; one over frequency and other over orientation.

For all of the image sets referred to above:

- Stage 1 presents the typical grey scale X-ray image with ROI's numbered for reference.
- Stage 2 presents the frequency decompositions for all the projections. The images are linked with their energy measures and the frequency of reconstruction.
- Stage 3 derives the distribution of grey levels for the frequency decompositions.
- Stage 4 derives the mean grey level per region of interest after reconstruction and these are presented for the whole range of frequency decompositions.
- Stage5 presents the mean grey level as an ensemble versus orientation and are shown for all frequencies.
- Stage 6 demonstrates the selected filter for all the orientations at maximal energy in comparison to its original image. There are selections of orientations at the maximal energy where the amount of detectability is altered by control of the orientation, in most of cases between 60 and 90 degrees. It is clear those frequencies in the range of f=0.3-0.8 are interesting for detection of liquids and organics.
- Stage 6.1 (Optional) Further expansion of this may be conducted when a 'YAUW' computational method is involved for more precise computation and detection
- Stage 7 shows the filtered image at maximal energy with wired edges over the filtered image at maximal energy filter reconstruction.
- Stage 7.1 (Optional). When an algorithm such as 'Eye Design' is involved in the computational process, additional image enhancement is achieved. An example is shown for biomolecules imaged by EM (chaperons) and detected.

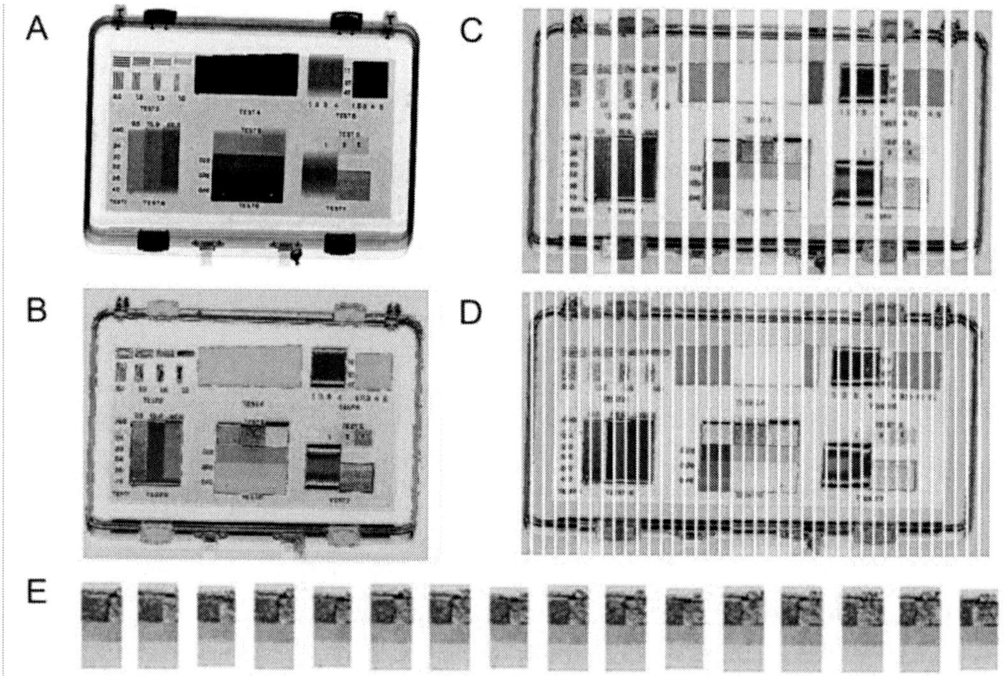

Figure 17. Steering algorithm results are presented. Here we show the effect on analysis over a portion of data acquired with a line detector as it is at airport scanners. A standard suitcase was imaged (image A) in dual energy mode (3D X-Ray Ltd) and presented in color-coding as follow: orange for organics, green for inorganics and blue for metals. The result of the analysis for the entire image is shown as image B in 'jet' colour coding. Images C and D show variation of the slit size on the analysed portion (also presented in E) of the image and its reconstruction. The simulated slit size is far below the 'physical demands' (slit width) of the real scanner, which shows a satisfactory result of the detection algorithms. Strip widths for image C,D and E are 50, 25 and 15 pixels respectively as published in 2015, Uroukov and Speller.

8.2. Directional Field (DF) Data Results

Much of the discussed concept is based around finding cues to suspicious materials such as explosives. Weapons also present characteristics that can be searched for. As discussed in the previous reports, DF is a technique, which is sensitive to 'a field's orientation' as may be seen around the edges of sharp test structures. We have shown that DF techniques are an alternative to detecting sharp objects that can also operate on small regions or image segments as might be found in a scanning system based upon the use of 1.5D imaging sensors, e.g., array segments made up of APS with a 512 x 32 pixels scanning system. This technique requires gradient computing that might be conducted using alternative filtering techniques such as local covariance.

8.3. Steering Algorithm

Extensive simulation tests were carried out to examine the minimal operational requirements of the I-ImaS concept when using the filter bank for building the feedback. The tests were made to see what was the minimal amount of information in the scout image that would be sufficient for effective operation of the filter bank. It was demonstrated that a scout image using a 32 pixel wide strip of data would provide the required image segment.Tests were conducted using strips or limited full field images as the scout image, as shown in Figure 17.

An 'optimal visibility concept' was found to be functional with the smaller area images even down to just a strip of 32 columns of data. Conditions where the object was passed through the slit detector were simulated and filter operations were computed with dropped and non-dropped lines.

Both slit width and image resolution were varied to see the effect on the filter operation. In all cases where the data width exceeded 32 columns satisfactory results were found from the filter bank.

8.4. Visibility Results

The effectiveness of the scanner concept was evaluated by looking at the enhanced visibility that the system could provide. The visibility model was tested for various ROI's, which correspond to materials. The sample was imaged at various tube currents (ranging from 10-500 μA) and potentials. The visibility values were computed for each ROI, i.e., per material. Clear results showed that the signal levels only change slowly, as the detected (filtered) signal per ROI remained the same. This demonstrates that the algorithms are robust, being based on textural coding rather than grey levels.

When the mean levels of the ROI's were plotted for a range of tube currents, they supported numerically the statement above by reminding in same range. The results show that the X-Ray intensity produces satisfactory measures of textures, achieving similar detection levels for all tube currents between 100 μA and 500 μA. The analysis does not support imaging conditions at 10 μA where the image is too contaminated by noise and the textures are lost in some ROI's.

It can be see that the visibility is not consistent for all ROIs, i.e., materials, and depends on the filter bank selectivity. This is associated with perceptibility that depends on the target object as defined in advance before the obscuring/observation process starts.

CONCLUSION

A novel approach to Intelligent Baggage Scanning was developed for security applications to enhance luggage checking and control at transportation facilities. It involves imaging processing and control steps and may be summarized as follow:

- A texture analysis method and supporting enhanced numerical model were developed to analyze X-ray imaged materials.
- The method employs a scout view image followed by an adapted control image and does not require dual energy imaging facilities.

- A mathematical model that describes the material properties of the imaged objects was developed (called 'Urukov's Code) and the use of this to segment the image into suspicious and normal regions was illustrated.
- As a further enhancement to the segmentation process, a Direction Field method was used to detect sharp objects (characteristics of many weapons) in luggage.
- A technique to enhance the detectability of selected materials using adapted control was developed through the use of a novel *Visibility model.* This model was inspired by human vision and was developed to control and enhance imaging on-the-fly.
- Using this adapted control visibility model we demonstrated between 200% and 500% signal enhancement for an ROI in the image.

The techniques described in this report can be implemented on existing hardware provided dual imaging (adaptable control) is available. It may be combined with existing software for analysis and provide specific searches (e.g., plastic and liquid explosives) which current scanners are unable to provide.

The variations of the image analysis were shown to be successfully operational over a range of imaging techniques such as medical ultrasound, electron microscopy, X-ray and X-ray phase contrast imaging and open an avenue for further exploration.

APPENDICES

Appendix 1. X-ray Phase Contrast Imaging and Uroukov's Code

The efficiency of the filtering approach and the model was demonstrated by tests on the X-ray phase contrast image (xPCI) in Figure 18. Here we show an enhancement of the contrast at particular image ROIs which include a 45 nm wire aligned in the blue square and titled with letter '1'.

Figure 18 Micro wire profiles obtained for a range of reconstructed frequencies. Signal profiles were extracted from a region of interest (roi) around a micro wire, indicated in the blue square labelled as '1' in panel A. The signal is obtained as the average over the lines in the roi, and plotted in absolute units versus pixel indexes along the roi as shown in B.

The level of the signal after filtering is measured as mean gray level per line averaging at the ROI. Figure 19 shows the enhanced contrast profiles for images, obtained at 40kVp and 100kVp, which are calculated at selected images (with maximal energy). The black line is the original profile and the red, pink and blue are 3 selected filter reconstructions where the profiles are obtained.

The enhancement for the selected ROI is between two and four times with respect to the original profile. Figure 20 demonstrates the original and selected (max energy) image.

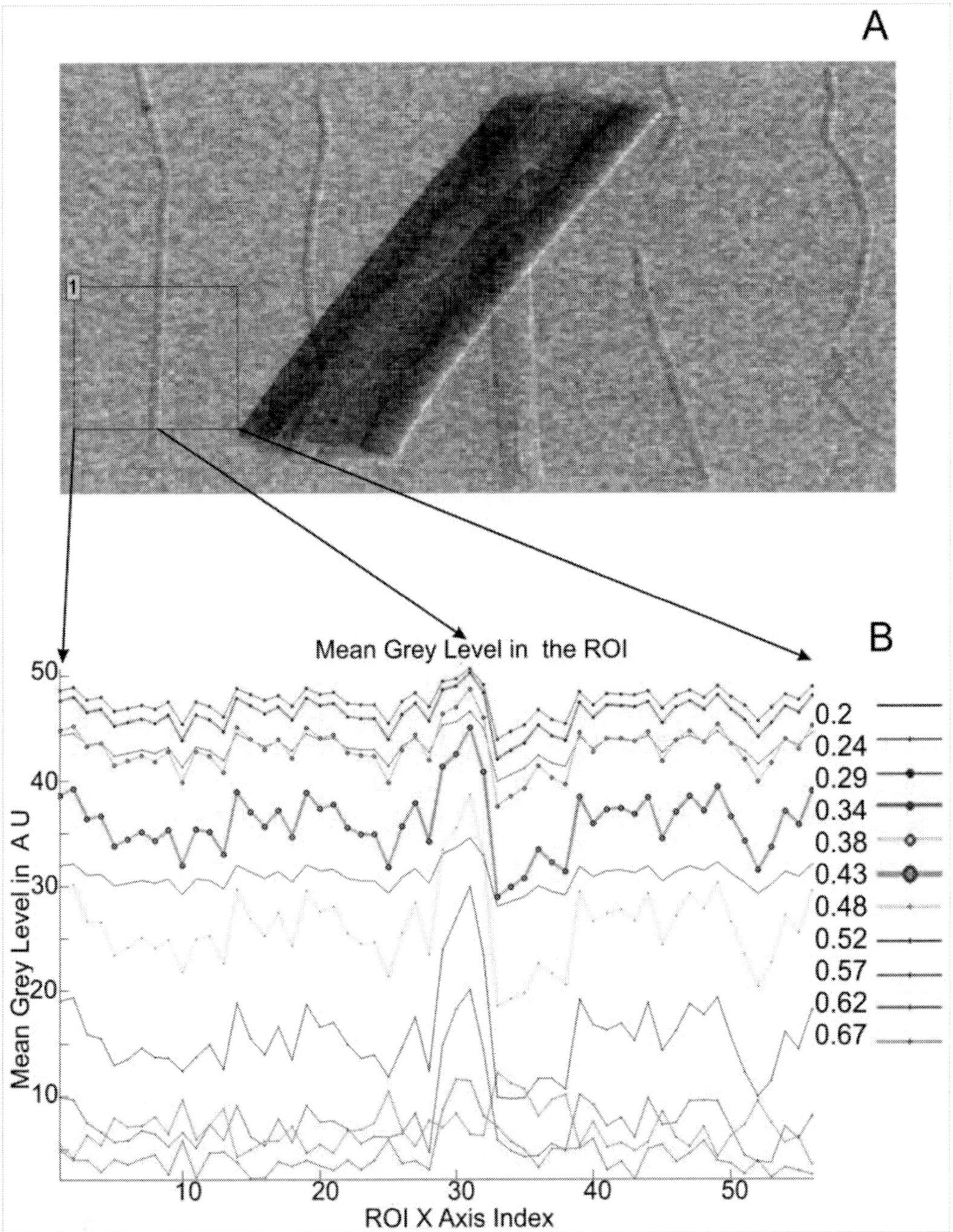

Figure 18. Microwire profiles.

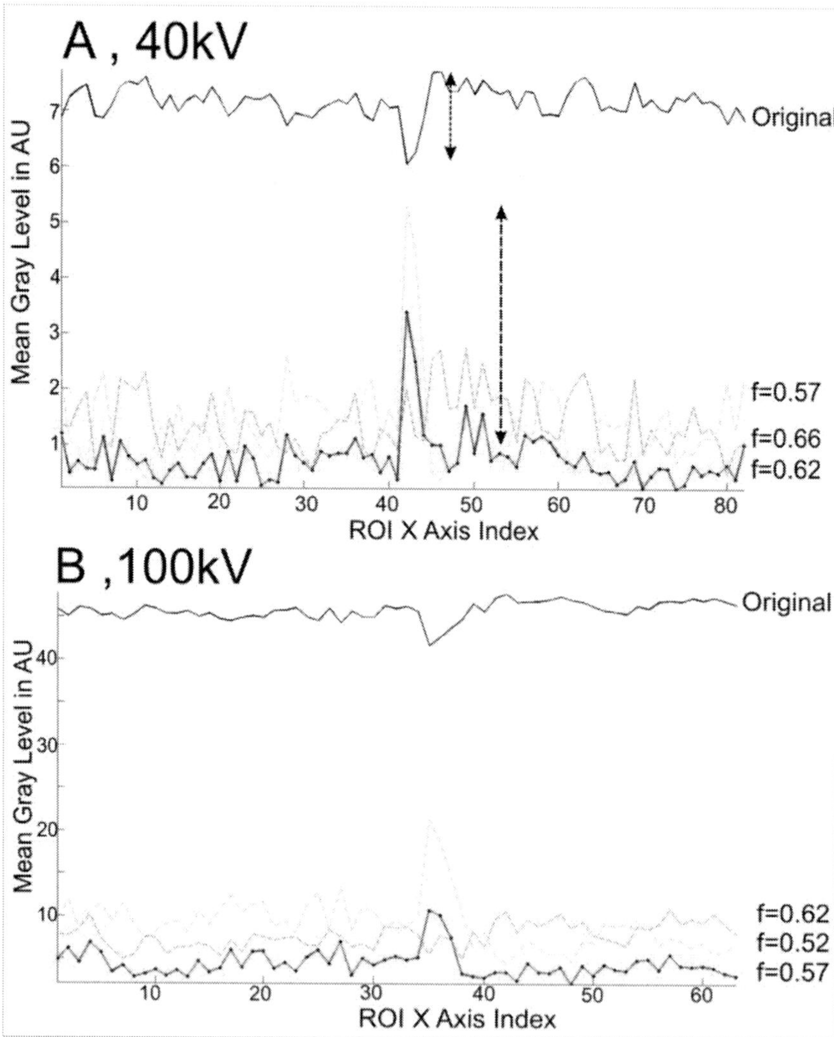

Figure 19. A comparison of contrast enhancement in the selected ROI, between processed and unprocessed images, under two different imaging conditions i.e., tube voltages of 40 kVp (A) and 100 kVp (B). The profiles obtained from the images are presented for a range of frequencies f. The frequency range is centred at the maximum energy image, and values one step below and one step above in the filterbank (f_{Emax}-1,f_{Emax},f_{Emax+1}) are also shown. In both A and B a contrast enhancement of the order of ~ 300% is observed.

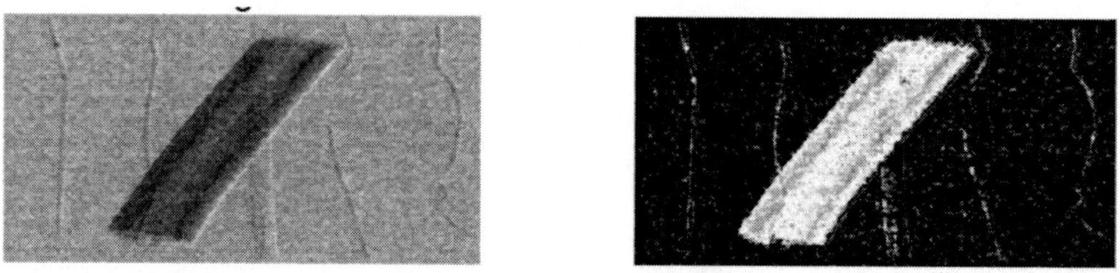

Figure. 20 A comparison of contrast enhancement for unprocessed and processed images acquired at a tube voltage of 40kV. The LHS shows the original image and the RHS the processed image with maximum energy measure.

Appendix 2. A Passengers Backpack. Tobacco Detection

In this example various organic products, liquids and tobacco were organised in layers, placed in a bag and imaged. The layers were combined and built as is described: *Layer 1* Organic materials comprising various foods. *Layer 2* Organic materials comprising cosmetic and liquid-gel substances. *Layer 3* Tobacco products.

With this example we demonstrate the result of two options, one to detect general organic substances and a second using specific tuning of the system to detect tobacco products when mixed with organics. Tobacco has been chosen as an example of a textured organic. Distinguishing tobacco from paper and other dry products with high fibre and cellulose content, based on their X-Ray absorption is a challenging problem. Standard systems fail to recognise similar products if they do not have spectroscopic detection. In this example we show that by suitable tuning of the filter bank it is possible to distinguish these materials.

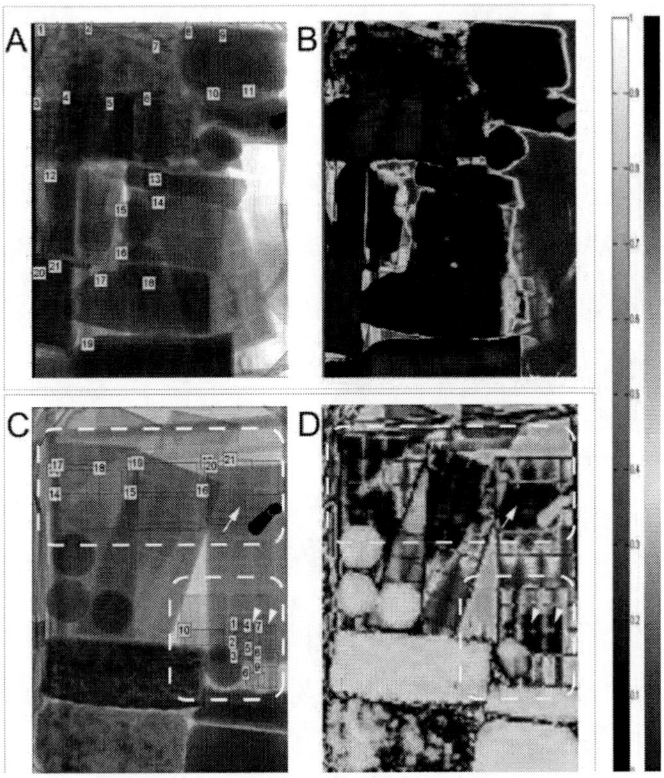

Figure 21. Organics and tobacco detection as published by Uroukov and Speller 2014. A panel shows an X-ray image of organic substancies. B panel shows a detection of the toal organic content in A. The white outliness in C and D indicate the locations of tobacco with arrows indicating loose tobbacco and arrowheads indicating boxed cigarettes tobacco.

Figure 21.A is the grey scale image on which are marked the ROIs with their index numbers. Figure 21.B presents the computed image at the selected maximum energy for a particular filter set. It shows the general detection of organic materials. Figure 21.C illustrates the specific localisation of tobacco as indicated by the white dashed boxes. Figure 21.D shows the results obtained using the tobacco detection filter set. Grey and jet colour scales are given for comparative purposes. On the X-ray image, the tobacco is detectable as the image is dominated by absorption from chocolate present in the bag. When specific detection is applied, the computed image shows the underlying hidden layers of tobacco.

Appendix 3. Medical Ultrasound Imaging

Transrectal ultrasound images were processed. Figure 22 shows the detection of prostate cancer foci. The images in Figure 22.A are titled with their filter index f1-f17. An original image is presented for comparative purposes titled 'Original'. Arrows pointing toward lesion localisation are shown on filter f13 and the original image. A prostate detection algorithm was designed for ultrasound imaging. The results are illustrated in Figure 22.B.

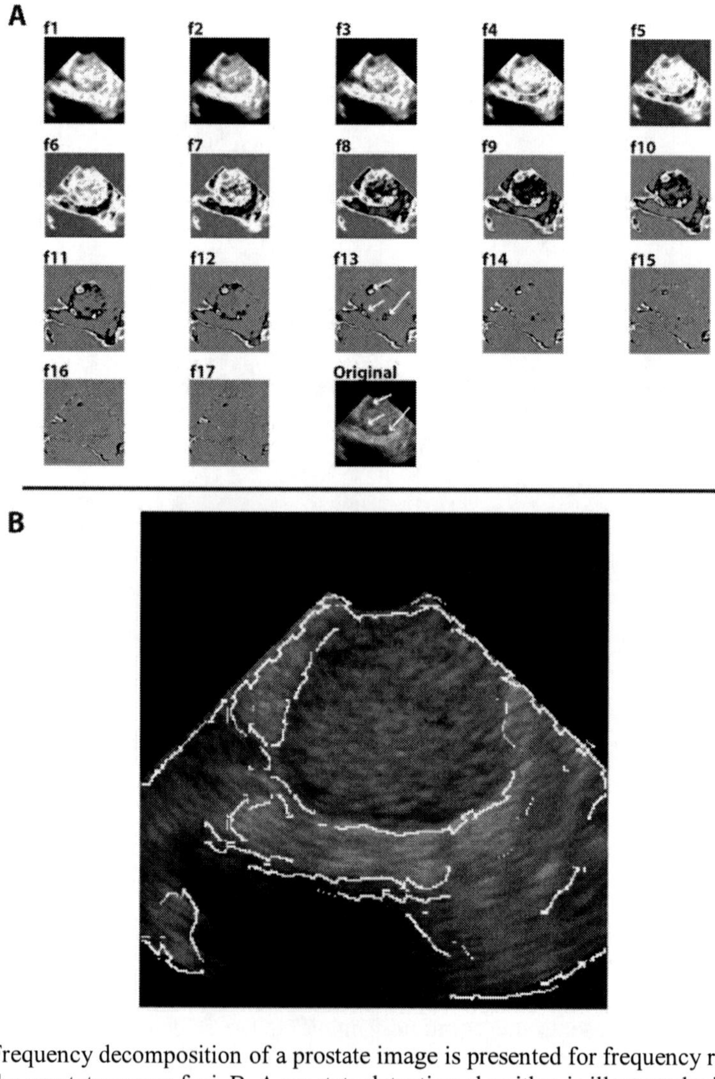

Figure 22. A. Frequency decomposition of a prostate image is presented for frequency ranges f_1-f17. Arrows show the prostate cancer foci. B. A prostate detection algorithm is illustrated with edge outlining applied over the original image. An artificial bias (outline shift) is used for better estimation of the organ's edge.

Appendix 4. Electron Microscopy Image Computing

Biological specimens such as viruses and cells may be imaged with electron microscopy. A low intensity beam of only several electrons per square angstrom is used to preserve the sample from electron beam damage. The images require filtering and contrast enhancement before they are further processed. The example shown here (Figure 23) is difficult to analyse using the existing software for object detection due to its very low signal to noise ratio. An algorithm 'Eye design' was applied to achieve a contrast level suitable for detection of the molecule complexes shown in Figure 23.

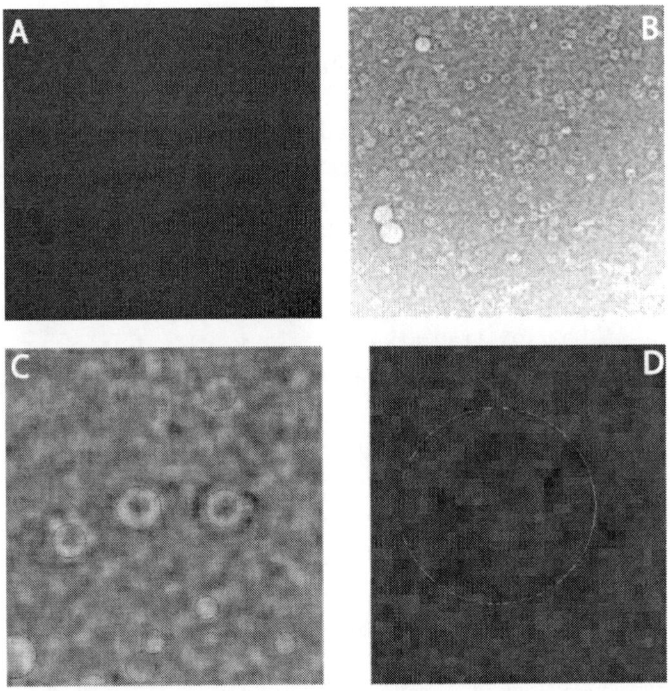

Figure 23. 'Eye Design' technique is applied to EM images of chaperone complexes for contrast enhancement and detection. A. A contrast scaled image. B. The image shown on A is enhanced using the 'Eye Design' algorithm. C. A zoomed image of part of (B) with chaperone complexes circled in red. D. An enlargement of a region of the original image shown in A with the location of a chaperone complex indicated by the red circle.

REFERENCES

AL-Hinnawi A. R. and Daear M., 2012. Image Texture Descriptors to Quantify Bilateral Filter on Low Dose Computerized Tomography. International Journal of Signal Processing. Image Processing and Pattern Recognition, 5(3), 123-136.

Campbell F.W. and Kulikowski J.J., 1966. Orientational sensitivity of human visual system. *Journal of Physiology*, 187, 437-455.

Clausi D. A., Jernigan M.E., 2000. Designing Gabor filters for optimal texture separability. *Pattern Recognition*, 33, 1835-1849.

Daugman J.G., 1985. Uncertainty relation for resolution in space, spatial frequency, and orientation optimized by two-dimensional visual cortex filters. *J. Opt. Soc. Am.* A.2(7), 1160-1169.

Gabor D. Theory of communication. 1945, 429-457.

Griffiths J. A., Metaxas M. G., Pani S., Schulerud H., Esbrand C., Royle G. J., Price B., RokvicT., LongoR., AsimidisA., Bletsas E., Cavouras D., Fant A., Gasiorek P., Georgiou H., Hall G., Jones J., Leaver J., LiG., Machin D., ManthosN., Matheson J., Noy M.,OstbyJ.M., Psomadellis F., van der SteltP.F., TheodoridisS., TriantisF., TurchettaR., VenanziC., SpellerR.D., 2008. Preliminary images from an adaptive imaging system, *Physica Medica*, 24(2), 117-21.

Huda W, Sajewicz. A. M., Ogden K.M., Dance D. R., 2003. Experimental investigation of the dose and image quality characteristics of a digital mammography imaging system. *Med. Phys.*, 30(3), 442-448.

Ivan S. Uroukov , Robert Speller , and Alessandro Olivo, 2014.Image Enhancement of X-ray Phase Contrast Images of Micro Objects. Application to the detection of small components in threat objects. *Signal Processing Research*, 2014, (3), 24-30.

Ivan S. Uroukov and Robert Speller. A Preliminary Approach to Intelligent X-ray Imaging for Baggage Inspection at Airports. *Signal Processing Research*, 2015,(4), 1-11.

Jain A. K. and Farrokhnia F., 1991. Unsupervised texture segmentation using Gabor Filters. *Pattern Recognition*, 24(12), 1167-1186.

Julesz B., 1962. Visual pattern discrimination. *IRE Trans. on Information Technology*, IT-8, 84-92.

Julesz B., 1981. A theory of preattentive texture discrimination based on first order statistics of textons. *Biol. Cybernetics*, 41, 131-178.

Julesz Bela, 2011. Preattentive human vision: link between neurophysiology and psychophysics. Chapter 14, AT & T Bell Labs, Published in: *Comprehensive Physiology*. DOI: 10.1002/cphy.cp010514.

Kass M., Witkin A., and Terzopoulos D., "Snakes: Active contour models," *Int. J. Comput. Vis., 1987, vol. 1, pp. 321-331.*

Lin Hong, Yifei Wan and Anil Jain, 1998. Fingerprint image enhancement: Algorithm and performance evaluation, *IEEE Transactions on pattern analysis and machine intelligence*.Vol 20,8, pp. 777-789.

Longo R., Asimidis A., Cavouras D., Esbrand C., Fant A., Gasiorek P., Georgiou H., Hall G., Jones J., Leaver J. Li G., Griffiths J., Machin D., Manthos N., Metaxas M., Noy M., Østby J. M., Psomadellis F., Rokvic T., Royle G., Schulerud H., Speller R., van der Stelt P.F., Theodoridis S., Triantis F., Turchetta R., Venanzi C., A scanning system for intelligent imaging: I-ImaS, Proc. SPIE 6510, Medical Imaging 2007: Physics of Medical Imaging, 65100M (March 13, 2007); doi:10.1117/12.708457, 2007.

Marcelja S., 1980. Mathematical description of the responses of simple cortical cells. *J. Opt. Soc. Am.*, 70(11), 1297-1300.

Martin C.J., Sutton D.G., Sharp P.F., 1999. Balancing patient dose and image quality. *Applied Radiation and Isotopes*, 50, 1-19.

Petkov N., Kruizinga P, 1997. Computational models of visual neurons specialized in detection of periodic and aperiodic oriented visual stimuli: bar and grating cells. *Biol Cybern.* 76,83-96.

Rao A. R. and Lohse G. L., 1993. Identifying high level features in texture perception. GVGIP: *Graphical Models and Image Processing*, 55 (3), 218-233.

About the Editor

Juan Zhang, PhD

Associate Professor
Key Laboratory of Specialty Fiber Optics and Optical Access Networks
School of Communication and Information Engineering
Shanghai University, Shanghai, China
Email: juanzhang@staff.shu.edu.cn

Juan Zhang received her PhD in 2004 from Shanghai Institute of Optics and fine Mechanics (SIOM), Chinese Academy of Sciences (CAS) and then joined school of communication and information engineering, Shanghai University, China where she now works as an associate professor at the key laboratory of specialty fiber optics and optical access networks. She worked as a visiting scholar at the University of Nottingham, UK, from 2011 to 2012. Her research fields include signal processing in optical communication, application of DSP technology in optics and photonics, design and application of optical interferometric devices, crystal optics and its applications etc. She is co-author of more than 60 scientific papers and 20 technical patents on these fields.

INDEX

N

O

P

R

S

T

U

V

W

X

Z

α